D0163536

Fresh Water Pollution I: Bacteriological, and Chemical Pollutants

A Volume in the MSS Topics in Ecology Series

Papers by
Charles W. Hendricks, Timothy B. Savage, James T. Staley, R. Jay Smith, E. H. Kampelmacher, Willard A. Bruce, A. Gene Collins, A. Benvenue, Jolin P. Haberman, K. H. Nelson, K. M. Grasshoff, John Cairns, Jr. et al.

MSS Information Corporation
655 Madison Avenue, New York, N.Y. 10021

Library of Congress Cataloging in Publication Data
Main entry under title:

Bacteriological & chemical pollution of fresh water.

(Fresh water pollution, 1)
1. Water--Pollution. 2. Bacteria, Pathogenic. 3. Chemicals. I. Hendricks, Charles W. II. Series. [DNLM: 1. Water pollution--Collected works. 2. Water pollution, Radioactive--Collected works. WA 689 F887]
TD420.F75 vol. 1   628.1'68'08s [628.1'68]
ISBN 0-8422-7076-0   72-13708

TABLE OF CONTENTS

CREDITS AND ACKNOWLEDGEMENTS

Bevenue, A.; T.W. Kelley; and J.W. Hylin, "Problems in Water Analysis for Pesticide Residues," *Journal of Chromatography*, 1971, 54:71-76.

Bruce, Willard A.; and Irving Grossman, "Oil — A New York State Pollution Problem," *Journal of the Water Pollution Federation*, 1971, 43:500-505.

Cairns, John, Jr.; Kenneth L. Dickson; and John S. Crossman, "The Response of Aquatic Communities to Spills of Hazardous Materials," *Proceedings of the 1972 National Conference of Hazardous Material Spills*, 179-197.

Collins, A. Gene, "Oil and Gas Wells — Potential Polluters of the Environment?" *Journal of the Water Pollution Control Federation*, 1971, 43:2383-2393.

Grasshoff, K.M.; and K.M. Chan, "An Automatic Method for the Determination of Hydrogen Sulphide in Natural Waters," *Analytica Chimica Acta*, 1971, 53:442-445.

Haberman, John P., "Polarographic Determination of Traces of Nitrilotriacetate in Water Samples," *Analytical Chemistry*, 1971, 43: 63-67.

Hendricks, Charles W., "Enteric Bacterial Metabolism of Stream Sediment Eluates," *Canadian Journal of Microbiology*, 1971, 17:551-556.

Kampelmacher, E.H.; and Lucretia M. van Noorle Jansen, "*Salmonella* — Its Presence in and Removal from a Wastewater System," *Journal of the Water Pollution Control Federation*, 1970, 42:2069-2070.

Kampelmacher, E.H.; and Lucretia M. van Noorle Jansen, "Reduction of *Salmonella* in Compost in a Hog-Fattening Farm Oxidation Vat," *Journal of the Water Pollution Control Federation*, 1971, 43:1541-1545.

Kampelmacher, E.H.; and Lucretia M. van Noorle Jansen, "Reduction of Bacteria in Sludge Treatment," *Journal of the Water Pollution Control Federation*, 1972, 44:309-313.

Nelson, K.H.; and I. Lysyj, "Analysis of Water for Molecular Hydrogen Cyanide," *Journal of the Water Pollution Control Federation*, 1971, 43:799-805.

Savage, Timothy B.; and Lewis P. Stratton, "Bacteriological Evaluation of Two Test Methods for Chlorine in Swimming Pools," *Applied Microbiology*, 1971, 22:809-811.

Smith, R. Jay; and Robert M. Twedt, "Natural Relationships of Indicator and Pathogenic Bacteria in Stream Waters," *Journal of the Water Pollution Control Federation*, 1971, 43:2200-2209.

Staley, James T., "Incidence of Prosthecate Bacteria in a Polluted Stream," *Applied Microbiology*, 1971, 22:496-502.

PREFACE

This three-volume collection on fresh water pollution is an addition to the MSS Topics in Ecology series.

Papers published from 1971-1972 are presented providing current information on new developments in major aspects of fresh water pollution including bacteriological, chemical, thermal and radioactive pollution. Application of the latest technical devices for the control of water pollution is also discussed.

Pollution of Water with Salmonella and Other Pathological Micro-organisms

Enteric bacterial metabolism of stream sediment eluates

Charles W. Hendricks

Introduction

The concern today by environmentalists with the concentration of nutrients in freshwater lakes and streams is largely the result of our increasing demands which are being placed upon this resource. For some time it has been known that terrestrial bacterial species can grow and reproduce in extremely dilute nutrient concentrations (4, 10, 12) of laboratory media, but most of these organisms are not involved in pathogenesis of man or higher animals. Enterobacteriaceae, however, not only contains bacteria which are indicators of fecal pollution, but others, such as *Salmonella*, *Shigella*, and *Arizona*, which can produce serious intestinal disease.

It is suspected that bottom sediments of lakes and streams play a major role in the recycling process of nutrients which allows for much of the observed heterotrophic growth (8), but the nature of the role is still quite vague. Studies by Malaney *et al.* (13) and Boyd and Boyd (3) indicate that sediments will stimulate the growth of bacterial species indigenous to freshwater lakes and streams. Work by Hendricks and Morrison (10), though, has shown that stream sediments have the capacity to bind basal nutrients loosely and that aqueous extracts of sediments will increase the rate of growth of various enteric species in high-quality water at 15°C and less. It was postulated by these investigators that this loosely bound material was probably available for microbial use within the natural environment.

The present investigation is primarily concerned with nutrient binding by river bottom sediments and conditions for its removal and use by enteric bacteria.

Materials and Methods

Study Sites

Water and bottom sediments were collected for investigation from the North Oconee River, a typical stream of the North Georgia piedmont in Clarke County, near Athens, Georgia. Three sites were selected for this study: one above the city where the water was free from urban contamination (site 1), another within the corporate limits but below the center of the city (site 2), and a third located about 750 meters below the municipal sewage facility whose effluent contributed a significant BOD to the stream (site 3).

Organisms

One strain each of *Escherichia coli*, ATCC 11775; *Enterobacter aerogenes* (*Aerobacter aerogenes*), ATCC 12658; *Proteus rettgeri*, *Arizona arizonae* (*Paracolobactrum arizonae*), *Shigella flexneri* A1, NCDC (Atlanta, Georgia); and *Salmonella senftenberg*, CPHS (Ottawa, Canada), was grown in Trypticase Soy Broth (BBL) at 30°C for 16 h. Cells were then harvested by centrifugation; washed 3 times in sterile, carbon-free, deionized water; incubated at 30°C for 4 h to expend endogenous metabolism; and rested at 4°C for 18 h before each experiment.

Experimental Substrates

River water and stream bottom sediments were collected from each study site. The river water was immediately sterilized in an autoclave at 121°C for 15 min and then frozen after samples were removed for chemical analysis. After collection, sediments from each site were divided into 30-g lots, and each aliquot was washed three consecutive times with 50-ml volumes of carbon-free deionized water. These slurries were agitated for 30 min with a magnetic mixer and then clarified by centrifugation. Each washing was collected, sterilized in the autoclave, and frozen for later use.

After determining optimal pH and buffer ionic strength for elution (Fig. 1), 30-g aliquots of washed sediment from each site were eluted with 50 ml of 0.3 *M* sodium phosphate buffer (5), pH 7.0, in separate experiments, and with river water from the site where the sediment was collected. These eluates were also autoclaved at

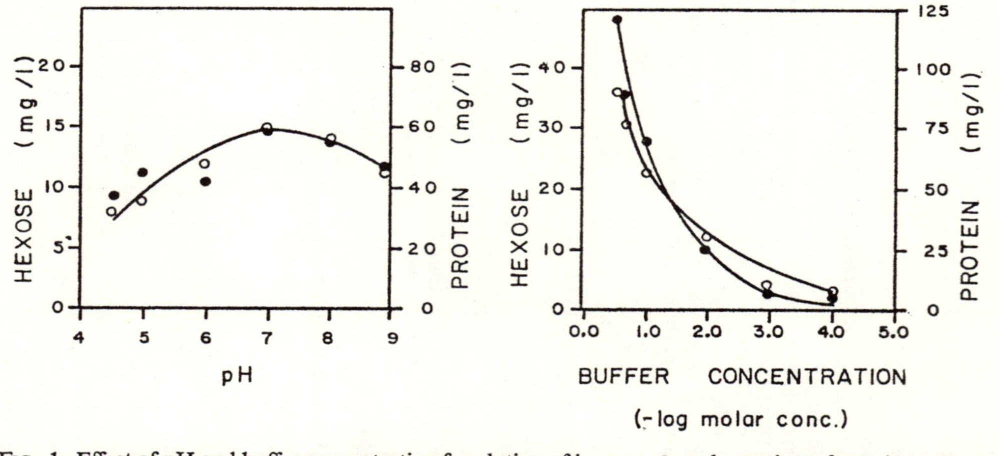

FIG. 1. Effect of pH and buffer concentration for elution of hexoses ○ and protein ● from river sediment. The pH was standardized to 7.0 in the studies to determine optimal buffer concentration for maximal elution.

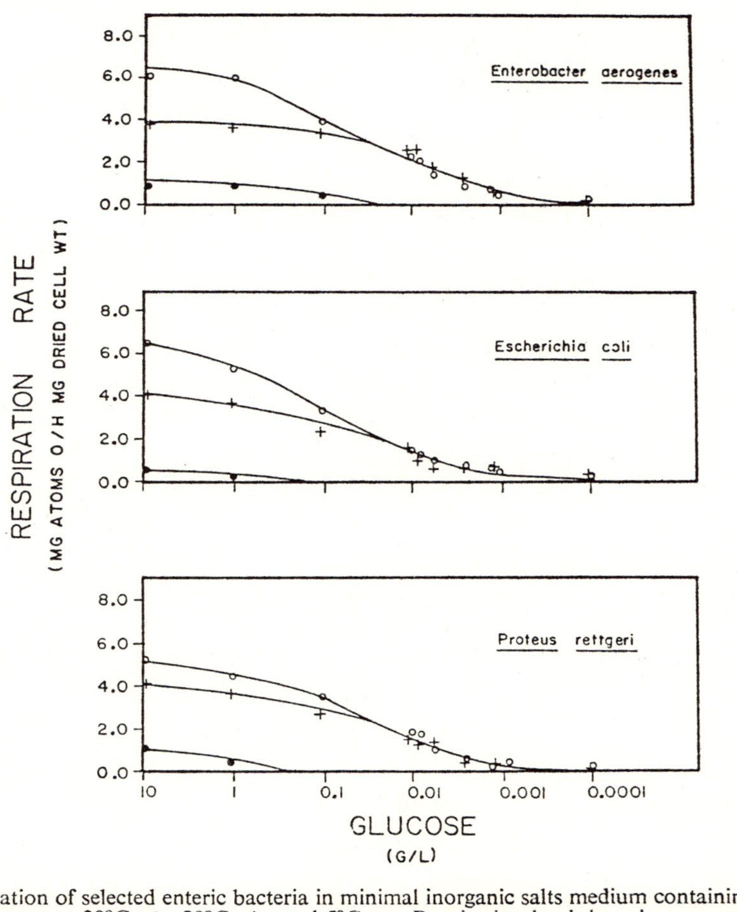

FIG. 2. Respiration of selected enteric bacteria in minimal inorganic salts medium containing various concentrations of glucose at 30°C, ○; 20°C, +; and 5°C, ●. Respiration levels have been corrected for endogenous activity.

121°C for 15 min and then frozen for later respiration studies and chemical analysis.

Analyses for nutritional constituents were made on the sediment washings and eluates as well as river water from each site by chemical procedures. Ammonia nitrogen and orthophosphate content was determined by procedures in *Standard Methods* (1), hexose was measured by anthrone (14), and protein content was determined by the Folin–Ciocalteau procedure (6).

Bacterial Respiration Studies with Eluted Sediments

Cell suspensions of each rested bacterial culture were prepared in deionized water and standardized to 0.9 optical density at 540 nm with a Spectronic 20 (Bausch and Lomb, Rochester, N.Y.) colorimeter-spectrophotometer. Dry cell weights were obtained by comparing the optical density with a standard curve which was prepared with organisms washed with deionized water and then dried overnight at 105°C.

Respiration studies were carried out with the use of a Biological Oxygen Monitoring System (Cole-Parmer, Chicago, Ill.) in which 4 ml of a particular substrate were placed into the monitor with 2 ml of the standardized bacterial suspension, and oxygen uptake was measured for 15 min. Individual experiments for each organism and substrate were run in duplicate at temperatures of 30°, 20°, and 5°C. Control substrates consisting of deionized water, river water from each site, varying dilutions of a minimal salts–glucose medium (7), and minimal medium containing 0.1, 0.2, and 0.3 *M* phosphate were run in each temperature series. Temperature of incubation was controlled within ± 0.2°C with a Lauda-Brinkman K-2/R Circulator (Westbury, N.Y.) and respiration was calculated as mg atoms O/h mg dried cell weight (17) after correcting for endogenous activity.

Results

Preliminary experiments indicated that 0.3 *M* phosphate buffer (pH 7.0) eluted the basal nutrients from the stream sediments in maximum concentration (Fig. 1). Table 1 presents data from experiments to determine the basic nutrient concentration of both the river water and extracts of sediments from each study site. Attempts to elute measurable quantities of the basal nutrients from the sediments with river water above those already present in the river water were without success.

Control experiments demonstrating basal respiration levels by the nonpathogenic enteric bacteria (*E. coli*, *E. aerogenes*, and *P. rettgeri*) with diluted minimal inorganic salts – glucose media are presented in Fig. 2. No uptake of oxygen was observed when the test organisms were run in deionized water before each respiration

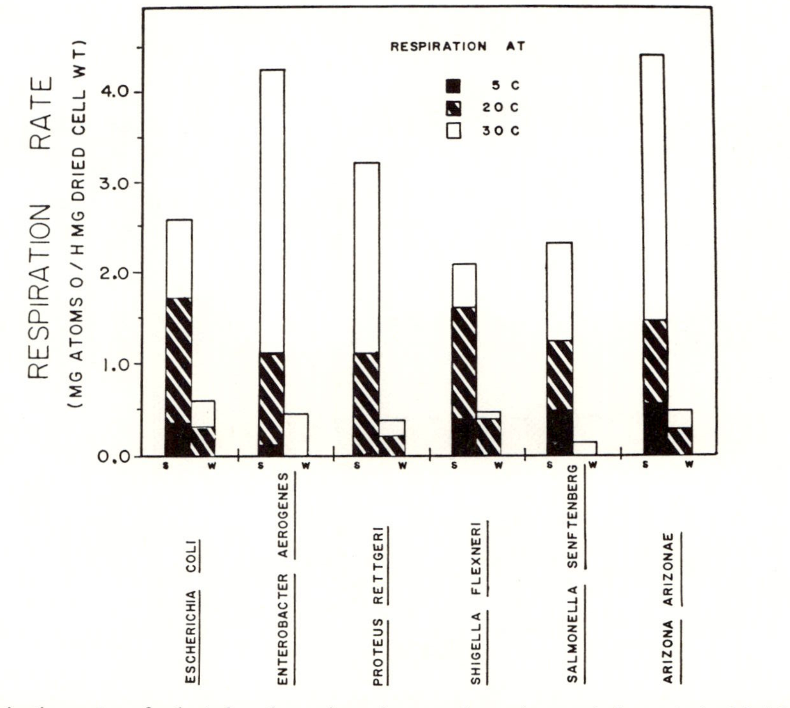

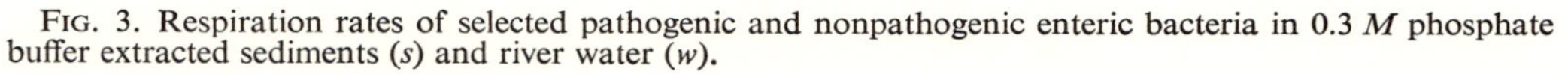
FIG. 3. Respiration rates of selected pathogenic and nonpathogenic enteric bacteria in 0.3 *M* phosphate buffer extracted sediments (*s*) and river water (*w*).

TABLE 1

Basic nutrient analysis of Oconee River water and extracts of river bottom sediment from the three study sites

Nutrient* assayed	Site	River water	Bottom sediment	
			Washed†	Buffer eluted‡
Ammonia nitrogen	1	1.6	0.6	9.0
	2	2.3	0.2	8.5
	3	4.7	1.0	25.0
Folin protein	1	13.6	1.5	45.0
	2	13.8	0.2	63.0
	3	18.3	0.2	120.0
Hexose	1	1.6	1.0	12.0
	2	2.8	0.6	22.5
	3	1.0	2.4	36.0
Orthophosphate	1	1.5	0.0	—§
	2	2.0	0.0	—
	3	4.1	0.4	—
pH	1	6.9	7.0	7.0
	2	7.0	7.0	7.0
	3	7.0	7.0	7.0

*Concentrations expressed in mg/l of sample.
†Concentration of nutrients present in the 3rd successive washing of deionized water.
‡Sediment eluted with 0.3 *M* phosphate buffer (pH 7.0).
§Phosphate concentrations were 0.3 *M*.

TABLE 2

Respiration of various enteric bacteria in Oconee River water and in extracts of river bottom sediments from site 1

Organism	Temp., C	River water*	Bottom sediment	
			Washed†	Buffer eluted‡
Escherichia coli	30	0.19§	0.34	1.98
	20	0.00	0.00	1.67
	5	0.00	0.00	0.44
Enterobacter aerogenes	30	0.22	0.46	1.60
	20	0.00	0.09	1.13
	5	0.00	0.16	0.00
Proteus rettgeri	30	0.26	0.12	1.64
	20	0.00	0.00	1.38
	5	0.00	0.00	0.00

*Respiration rates expressed as mg atoms oxygen (O)/h mg dry cell weight.
†After third successive wash.
‡0.3 *M* phosphate buffer (pH 7.0).
§All respiration rates have been corrected for endogenous activity.

TABLE 3

Respiration of various enteric bacteria in Oconee River water and in extracts of river bottom sediments from site 2

Organism	Temp., C	River water*	Bottom sediment	
			Washed†	Buffer eluted‡
Escherichia coli	30	0.43§	0.23	1.81
	20	0.21	0.52	1.68
	5	0.00	0.00	0.18
Enterobacter aerogenes	30	0.53	0.37	1.26
	20	0.00	0.00	1.27
	5	0.00	0.24	0.10
Proteus rettgeri	30	0.25	0.00	1.68
	20	0.00	0.21	0.80
	5	0.00	0.00	0.00

*Respiration rates expressed as mg atoms oxygen (O)/h mg dry cell weight.
†After third successive wash.
‡0.3 *M* phosphate buffer (pH 7.0).
§All respiration rates have been corrected for endogenous activity.

TABLE 4

Respiration of various enteric bacteria in Oconee River water and in extracts of river bottom sediments from site 3

Organism	Temp., C	River water*	Bottom sediment	
			Washed†	Buffer eluted‡
Escherichia coli	30	0.58§	0.22	2.58
	20	0.33	0.10	1.73
	5	0.00	0.00	0.34
Enterobacter aerogenes	30	0.45	1.34	4.25
	20	0.00	0.90	1.13
	5	0.00	0.00	0.10
Proteus rettgeri	30	0.36	0.36	3.23
	20	0.17	0.12	1.07
	5	0.00	0.00	0.00

*Respiration rates expressed as mg atoms oxygen (O)/h mg dry cell weight.
†After third successive wash.
‡0.3 *M* phosphate buffer (pH 7.0).
§All respiration rates have been corrected for endogenous activity.

rate determination, and no apparent phosphate effect upon respiration was observed at the final experimental concentration level of 0.2 *M*.

Use of the substrates present in the river water and in washed and buffer eluted bottom sediments is shown in Tables 2 through 4 and in Fig. 3. Tables 2, 3, and 4 express the respiration rates of the nonpathogenic enterics at 30°, 20°, and 5°C with the test substrates from sites 1, 2, and 3 respectively. In Fig. 2 are compared the respiration rates of the nonpathogenic enterics with pathogenic species of *S. flexneri*, *S. senftenberg*, and *A. arizonae* in buffer eluted sediments and river water from below the sewage plant (site 3).

Discussion

Previous studies (10, 12) have indicated that basal nutrients in concentrations about equal to Davis's (7) minimal salts – glucose medium diluted 1:1000 were sufficient for the maintenance and limited growth of prototrophic enteric bacteria. Data obtained in this study (Table 1) demonstrated that the ammonia nitrogen, carbon, and orthophosphate present in the three test substrates were comparable in concentration to those media prepared in the laboratory (4, 12). With the exception of hexose present in the river water, concentrations of the basal nutrients contained in both the river water and buffer-extracted bottom sediments were maximal at the site located below the sewage outfall (site 3).

Table 1 also demonstrates that bottom sediments from the Oconee River can be washed relatively free from loosely associated material and that a very high concentration of ammonia nitrogen, protein, and hexose can be sorbed onto the river bottom sediments. This observation becomes even more significant because three successive washings with deionized water and trial elution with river water could not remove the tightly bound material. Elution, though, was accomplished with buffer of ionic strengths which might be common to only the most severely polluted aquatic environments. These data suggest that the basal nutrients were very tightly adsorbed on the sands and clays forming the stream bottom sediments, and that they may not be readily available for metabolism by aquatic microorganisms. We have tested this concept with BOD studies in our laboratory and found that washed sediments had no stimulating effect on the oxygen demand. This is in agreement with Weber and Coble (20), who have found that cationic pesticides which were subject to microbial degradation could be adsorbed on various clays and were then no longer subject to decomposition or even readily available for plant uptake.

Respiration of organisms in the aquatic environment has been used as a means of estimating *in situ* activity (16, 18, 19). Control experiments (Fig. 2) suggest that a respiration rate of about 0.5–3.5 mg atoms O/h mg dried cell weight could be achieved if the carbon (hexose and protein) analyses of the river water and extract bottom sediments represent readily oxidizable substrates. Similar rates would be expected with the prepared natural substrates at both the 30° and 20°C incubation temperatures, but respiration at 5°C should be minimal or nonexistent. Results of the respiration rate studies using river water and extracted sediments confirmed this hypothesis (Tables 2, 3, and 4) and reflected the basal nutrient concentration. Respiration rates above endogenous levels for all organisms tested in river water were lowest at site 1, but as nutrient concentration increased in the water from sites 2 and 3, respiration rates at both 20° and 30°C approached the predicted values (Tables 3 and 4). Sediment eluted with phosphate buffer in all cases yielded respiration rates far exceeding those observed with river water. With the exception of *Enterobacter aerogenes* at site 1, both *Escherichia coli* and *Enterobacter aerogenes* could use the substrates present in the eluates from all sites at 5°C. Buffer eluted sediments from below the sewage plant demonstrated the highest respiration rates achieved at 30°C with rates at 20° and 5°C approximating those at the other two sites. When these rates were compared with those for the pathogenic species (*Shigella flexneri*, *Salmonella senftenberg*, and *Arizona arizonae*), equivalent respiration was observed (Fig. 3). These data indicate that both pathogenic as well as nonpathogenic bacteria could use substrates that were present in the river water and those adsorbed on the bottom sediments after relatively mild laboratory treatment.

Although these data cannot be directly extrapolated to give information on the fresh water ecosystem, a model can be proposed which suggests bottom sediments can control the nutrient material which is suspended in the water. Although Wright and Hobbie (21) have shown that the rate at which nutrients turn over to be significant for sustaining heterotrophic bacterial growth in lake water, Hendricks and Morrison (9, 10) have speculated that only a minor component of the total bacterial growth and reproduction can probably occur within free-flowing water of a high-quality stream since nutrients are normally in low concentration. However, estimations of nutrient concentration and numbers of organisms are not constant within fresh or marine environments and may vary considerably from time to time (2, 11, 15). These observations can be explained on the basis that a portion of the sediment–nutrient complex can be removed by aqueous extraction which could then be resuspended or dissolved in the water. As the apparent nutrient concentration increased in the river water, increased heterotrophic bacterial growth might also be observed. However, once the adsorptive capacity of the sediments has been reached, as perhaps exists around sewage plant effluents, stream nutrients then could not be removed from the system and much growth of aquatic organisms could result. An occurrence such as this would seriously affect the aerobic component of the system and lead to an altered self-purification potential for some distance below a sewage outfall.

Acknowledgments

This investigation was supported by a research grant 16050 EQS from the Federal Water Quality Administration, U.S. Department of the Interior. I thank Gay Walter for her valuable technical assistance.

1. American Public Health Association. 1966. Standard methods for the examination of water and waste

water. 12th ed. American Public Health Association Inc., New York.

2. Anthony, E. H. 1970. Bacteria in estuarine (Bras d'Or Lake) sediment. Can. J. Microbiol. **16**: 373–389.
3. Boyd, W. L., and W. Boyd. 1962. Viability of thermophiles and coliform bacteria in arctic soils and water. Can. J. Microbiol. **8**: 189–192.
4. Butterfield, C. T. 1929. Experimental studies of natural purification in polluted waters. III. A note on the relation between food concentration in liquid media and bacterial growth. Public Health Rep. (U.S.). **44**: 2865–2872.
5. Colowick, S. P., and N. O. Kaplan. 1955. Methods in enzymology. Vol. 1. Academic Press, Inc., New York.
6. Colowick, S. P., and N. O. Kaplan. 1955. Methods in enzymology. Vol. 3. Academic Press, Inc., New York.
7. Davis, B. D. 1950. Nonfiltrability of the agents of genetic recombination in *Escherichia coli*. J. Bacteriol. **60**: 507–508.
8. Harter, R. D. 1968. Adsorption of pnosphorus by lake sediment. Soil Sci. Soc. Amer. Proc. **32**: 514–518.
9. Hendricks, C. W., and S. M. Morrison. 1967. Strain alteration in enteric bacteria isolated from river water. Can. J. Microbiol. **13**: 271–277.
10. Hendricks, C. W., and S. M. Morrison. 1967. Multiplication and growth of selected enteric bacteria in clear mountain stream water. Water Res. **1**: 567–576.
11. Low, P. F., B. G. Davey, K. W. Lee, and D. E. Baker. 1968. Clay sols versus clay gels: biological activity compared. Science, **161**: 897.
12. McGrew, S. B., and M. F. Mallette. 1962. Energy of maintenance in *Escherichia coli*. J. Bacteriol. **83**: 844–850.
13. Malaney, G. W., H. H. Weiser, R. O. Turner, and M. Van Horn. 1962. Coliforms, enterococci, thermodurics, thermophiles and psychrophiles in untreated farm pond waters. Appl. Microbiol. **10**: 44–51.
14. Morris, D. L. 1948. Quantitative determination of carbohydrates with Dreywood's anthrone reagent. Science, **104**: 254–255.
15. Morrison, S. M., and J. F. Fair. 1966. Influence of environment on stream microbial dynamics. Hydrology paper No. 13. Colorado State Univ., Fort Collins, Colorado.
16. Olson, T. A., and M. E. Rueger. 1968. Relationship of oxygen requirements to index-organism classification of immature aquatic insects. J. Water Pollut. Contr. Fed. **40**: (Res. Suppl.) R188-R202.
17. Pomeroy, L. R., and R. E. Johannes. 1968. Occurrence and respiration of ultraplankton in the upper 500 meters of the ocean. Deep-Sea Res. **15**: 381–391.
18. Rueger, M. E., T. A. Olson, and J. L. Scofield. 1968. Oxygen requirements of benthic insects as determined by manometric and polarographic techniques. Water Res. **3**: 99–120.
19. Schroeder, E. D. 1968. Importance of the BOD plateau. Water Res. **2**: 803–809.
20. Weber, J. B., and H. D. Coble. 1968. Microbial decomposition of diquat adsorbed on montmorillonite and kaolinite clays. J. Agr. Food Chem. **16**: 475–478.

21. WRIGHT, R. T., and J. E. HOBBIE. 1965. The uptake of organic solutes in lake water. Limnol. Oceanogr. **10**: 22–28.

Bacteriological Evaluation of Two Test Methods for Chlorine in Swimming Pools

TIMOTHY B. SAVAGE AND LEWIS P. STRATTON

Chlorine is commonly used to control bacteria in swimming pools. Free chlorine atoms kill bacteria by chlorinating outer components of the cell and breaking peptide bonds as do other halogens (5), but chlorine bound to protein or amines loses its effectiveness to a large extent (2). It is difficult to determine accurately the level of free chlorine ions in pool water. The standard method of determining chlorine concentrations in pools is the orthotolidine method, which is not selective for free chlorine, but registers bound chlorine as well.

In 1970 a swimming pool water-testing device utilizing indicators was developed as a test for free chlorine and *p*H (C. O. Rupe et al., Abstr. no. 6, 160th Amer. Chem. Soc. Nat. Meet., 1970). These test strips (Aqua Check; Ames Co., Division of Miles Laboratory, Inc., Elkhart, Ind.) consist of a plastic rectangle (89 by 25 mm) containing absorbent paper treated with a mixture of syringaldazine and vallinazine, which turns from yellow to purple in the presence of free chlorine, and a *p*H-indicator strip.

This test (syringaldazine and vanillinazine; SV) has been subjected to various laboratory and field tests concerning its chemial accuracy by several workers.

It was found that the SV test gave approximately the same results as standard chemical tests for free chlorine but different results than did orthotolidine when bound chlorine was tested (4; Rupe et al., Abstr. no. 6, Amer. Chem. Soc. Nat. Meet., 1970). However, bactericidal properties of swimming pool water were not tested.

This paper summarizes studies which compared the ability of the orthotolidine and SV test methods to detect potentially dangerous conditions in pools and their relative sensitivities to changes in free chlorine levels.

MATERIALS AND METHODS

Three common bacterial species of medical importance were used in this study. *Streptococcus faecalis* is usually considered to be one of the best indicators of human fecal contamination of water systems. *Escherichia coli* has long been considered the standard for indication of fecal contamination of water supplies. *Pseudomonas aeruginosa* is found in ear infections and some external wounds and is thus of considerable interest in determining the safety of pool water (G. A. Pottz, personal communication). Wild-type strains were used.

Stock cultures of bacteria were maintained on nutrient agar (Difco) and transferred every 3 weeks.

The orthotolidine test was performed by using a Guardex Pool Test Kit, according to the manufacturer's directions (3).

M-Endo Broth MF (Difco) was used as a preferential medium for *E. coli*, Streptococcus faecalis (SF) Broth (Difco) was the preferential medium for *S. faecalis*, and nutrient broth (Difco) was used for *P. aeruginosa* since no preferential medium was available. Absorbent pads (50 mm) and membrane filters (0.45-μm pore size) were obtained from Schleicher and Schuell. All incubations were carried out at 37 C.

Swimming pool water from a community pool was obtained from a depth of 6 to 12 inches (ca. 15.24 to 30.48 cm) below the surface, autoclaved, and used immediately.

Erlenmeyer flasks (500 ml) with 100 ml of water were used to test for bacterial survival. Water was placed in the flasks and sterilized; sodium hypochlorite solution prepared by an approximately 50-fold dilution of Chlorox (Chlorox Co., Oakland, Calif.) was added until the SV test indicated 1.0 to 2.0 μg (ppm) of chlorine per ml which is considered a safe level for pools. The sodium hypochlorite solution was calibrated by the SV test and orthotolidine to insure uniform conditions before each set of flasks was tested.

TABLE 1. *Chlorine levels and Escherichia coli survival in swimming pool water, after addition of human urine*

Sample	Urine added[a]	Chlorine indicated (μg/ml)				Bacteria[d]
		Before urine		After urine		
		SV[b]	OT[c]	SV	OT	
1	0	1.0	1.0	1.0	1.0	0
2	0	1.0	1.0	1.0	1.0	0
3	0.05	1.5–2.0	1.5	1.5	1.5	0
4	0.05	1.5	1.5	1.5	1.5	0
5	0.15	1.5–2.0	1.5	0	1.5	640
6	0.25	1.5	1.5	0	1.5	1,000

[a] Milliliters of fresh human urine added per 100 ml of water after adjustment of chlorine content.
[b] SV test strips.
[c] Orthotolidine pool test kit.
[d] Viable bacteria per 100-ml test flask after equilibration, determined by membrane filter technique. (About 5×10^5 cells were added to each test flask.)

TABLE 2. *Chlorine levels and Pseudomonas aeruginosa survival in swimming pool water after addition of human urine*

Sample	Urine added[a]	Chlorine indicated (μg/ml)				Bacteria
		Before urine		After urine		
		SV	OT	SV	OT	
1	0	1.0	1.0	1.0	1.0	0
2	0	1.0	1.0	1.0	1.0	0
3	0.05	1.5–2.0	1.5	1.0	1.5	0
4	0.05	1.5	1.5	1.0	1.5	0
5	0.15	1.5	1.5	0	1.5	520
6	0.25	2.0	1.5	0	1.5	2,700

[a] *See* footnotes for Table 1.

TABLE 3. *Chlorine levels and Streptococcus faecalis survival in swimming pool water after the addition of human urine*

Sample	Urine added[a]	Chlorine indicated (μg/ml)				Bacteria
		Before urine		After urine		
		SV	OT	SV	OT	
1	0	2.0	1.5	2.0	1.5	0
2	0	2.0	1.5	2.0	1.5	0
3	0.05	2.0	1.5	1.0	1.5	0
4	0.05	2.0	1.5	1.0	1.5	0
5	0.15	2.0	1.5	0	1.5	250
6	0.25	2.0	1.5	0	1.5	1,600

[a] *See* footnotes for Table 1.

Urine and the bacteria to be tested (0.005 ml of an overnight culture with 10^8 to 2×10^8 cells per ml) were added sequentially to the flasks with omissions as indicated in the tables. Flasks were mixed and allowed to equilibrate for 15 min after each addition.

After final equilibration, the number of viable bacteria was determined by the membrane filter technique.

RESULTS AND DISCUSSION

When sodium hypochlorite solution alone was added to the flasks, the test results of orthotoli-

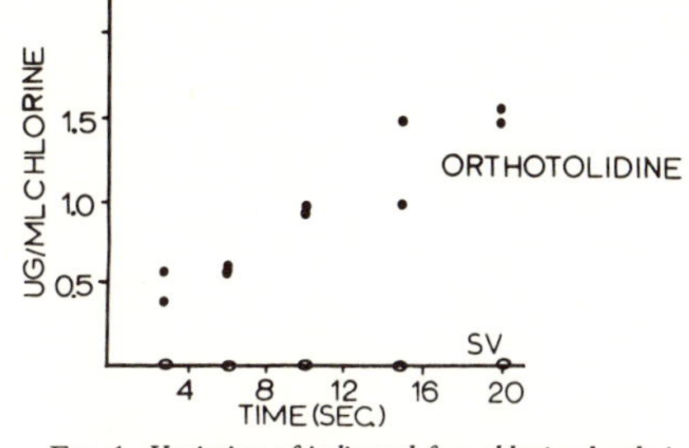

FIG. 1. *Variation of indicated free chlorine levels in the presence of bound chlorine with time for orthotolidine and SV test methods.*

dine and the SV test showed good agreement. It appeared that when only free chlorine was present the test results of these two methods were consistent.

Urine and bacteria contain amines which bind chlorine and lower its effective concentration. When testing with SV, it was found that the initial chlorine level of 1.5 μg/ml dropped after the addition of 0.05 to 0.25 ml of urine to flasks containing 100 ml of water (Table 1–3). Similar tests with the orthotolidine method showed little or no change in the indicated level of free chlorine after addition of the urine. Viable *E. coli* in these samples ranged from 640 to 1,000 per 100 ml (Table 1). Under the same condition, *P. aeruginosa* (Table 2) varied from 520 to 2,700 bacteria per 100 ml, and *S. faecalis* ranged from 250 to 1,600 colonies (Table 3).

With a larger (0.1 ml) inoculation of an overnight culture (10^8 to 2×10^8 cells per ml) of *E. coli*, the solution had final SV test readings of zero and final orthotolidine readings of approximately 1.5 μg per ml of chlorine. These samples were found to contain from 2.0×10^6 to 1.8×10^7 viable *E. coli* per 100 ml.

In every instance in which SV gave a "safe" reading, there were no viable bacteria detectable by growth on M-Endo, SF, or nutrient broth. However, in all instances of "unsafe" SV readings and "safe" orthotolidine readings shown in Table 1–3, the bacteria counts per 100 ml were above the standards of the South Carolina Board of Health (four coliforms per 100 ml).

Similar tests were carried out starting with distilled water. The results obtained were identical although more sodium hypochlorite solution had to be added initially to obtain the necessary 1.0 to 2.0 μg of free chlorine per ml level. This is because the swimming pool water already had some active chlorine.

It has been reported by Black et al. (1) that orthotolidine testing must be done at 0 to 1 C and within 5 sec to obtain accurate free chlorine results in the presence of chloramines. Due to the difficulty of obtaining readings within this time interval under normal pool testing conditions, we investigated the effects of time on the results obtained. These are shown in Fig. 1, and indicate that readings at somewhat longer times (10 to 20 sec) give inaccurate results under conditions where bound chlorine is present (0.25 ml of fresh urine added at zero time).

The SV test provides a more precise measurement of effective antimicrobial activity of chlorinated water than does the orthotolidine test method commonly used for swimming pool testing. Orthotolidine fails to detect possible dangerous situations which are easily found with the SV test.

ACKNOWLEDGMENTS

We thank C. Bradley Hager, Ames Co., Miles Laboratories, Inc., Elkhart, Ind., for providing the Aqua Check test strips.

LITERATURE CITED

1. Black, A. P., M. A. Kerrin, J. J. Smith, G. M. Dykes, and W. E. Harlan. 1970. The disinfection of swimming pool water. II. A field study of the disinfection of public swimming pools. J. Pub. Health **60**:740–750.
2. Fetner, R. H., and R. S. Ingols. 1956. A comparison of the bactericidal activity of ozone and chlorine against *Escherichia coli* at 1°. J. Gen. Microbiol. **15**:381–385.
3. Field, C. A. Guardex instruction booklet. Purex Corporation, Ltd.
4. Geeting, D. G., C. B. Hager, and A. H. Free. 1970. Aqua check—a swimming pool information system. A bulletin of Ames Technical Services Laboratory, Ames Co., Division of Miles Laboratories, Inc., Elkhart, Ind.
5. McKee, J. E., C. J. Brokaw, and R. T. McLaughlin. 1960. Chemical and colocidal effects of halogens in sewage. J. Water Pollut. Control Fed. **32**:795–819.

Incidence of Prosthecate Bacteria in a Polluted Stream

JAMES T. STALEY

Unlike typical unicellular bacteria, each genus of heterotrophic, prosthecate bacteria contains individuals that are morphologically distinctive and recognizable. Based on the number, size, shape, and location of their appendages and the morphology of their cells, certain individuals can be identified as to genus and in some cases even species by phase microscopic examination of a population. Thus, the genus *Caulobacter* contains stalked cells that have a single thin prostheca extending from one pole of the cell (9). The genus *Asticcacaulis* is identical except its prostheca is borne in a subpolar position (9). *Hyphomicrobium* and *Pedomicrobium* produce buds at the distal tips of their wider prosthecae, a feature that distinguishes them from all other genera of heterotrophs. *Pedomicrobium* may produce several appendages from any location on the cell surface (2), whereas *Hyphomicrobium* normally produces only polar appendages (7). The other genera of heterotrophic prosthecate bacteria, *Prosthecomicrobium* and *Ancalomicrobium*, have cells with several appendages, but neither genus produces prosthecal buds. The appendages on *Ancalomicrobium* are fewer and longer than those of *Prosthecomicrobium* (10).

In addition to these heterotrophic forms, there are two genera of photosynthetic prosthecate bacteria, *Rhodomicrobium* (4) and *Prosthecochloris* (5). Unlike those mentioned previously, these are obligately anaerobic and frequently occur in multicellular aggregates. These phototrophic genera contain cells that are morphologically indistinguishable from heterotrophic counterparts, namely, *Hyphomicrobium* and *Prosthecomicrobium*, respectively, although the pigmentation of individuals may occasionally be intense enough to permit direct microscopic identification.

Because the prosthecate bacteria can be recognized by phase microscopic examination, they are ideal unicellular prokaryotes for direct identification and quantitative enumeration in the normal microflora of the natural environments in which they are found. There are, however, two sources of error in this approach. For one, not all of the individuals of a population may be identifiable. For example, swarmer cells in the genera *Caulobacter*, *Asticcacaulis*, and *Hyphomicrobium* would appear as rods or vibrios before the development of noticeable prosthecae. This would lead to an underestimation of their numbers. Another source of error involves mistaken identity. Stalked caulobacters, for instance, appear very much like some nonbudding stages of *Hyphomicrobium* and might be identified incorrectly. Similarly, individual cells of the photosynthetic forms could be mistaken for their heterotrophic counterparts as suggested previously. Despite these limitations, direct enumeration studies of these bacteria should be undertaken to provide some knowledge of their numerical significance in their natural habitats.

An alternative approach is quantitative viable counting. Belyaev (3) was the first to use the extinction-dilution technique for enumerating prosthecate bacteria when he determined the numbers of caulobacters in Russian reservoirs. In this procedure, replicate 10-fold serial dilutions of water samples are made in a liquid medium that will support the growth of heterotrophic prosthecate bacteria. After incubation, wet mounts from each tube are examined for the presence of representative forms. When a form distinctive to a genus is found, the tube is recorded positive for the genus.

By use of both direct and viable enumeration procedures, a study was conducted to determine the incidence of heterotrophic prosthecate bacteria in a polluted stream in Michigan. An attempt was also made to correlate the incidence of prosthecate bacteria to the coliform organisms along the watercourse.

MATERIALS AND METHODS

Study stream. The Red Cedar River is a warm-water stream that drains 489 square miles in south central Michigan. It flows some 40 miles in a northwesterly direction from its source in Cedar Lake to its confluence with the Grand River in Lansing (Fig. 1). Nine sampling sites were chosen along the watercourse, the first located upstream of Fowlerville at the Van Buren bridge. The second sampling site was the Gregory Road bridge downstream of Fowlerville and its sewage treatment plant. Sites 3 and 4 were located at Gramer Road bridge and Dietz Road bridge, respectively. This upstream section of the stream has a mud and silt bottom.

About 2 miles downstream of the Dietz Road

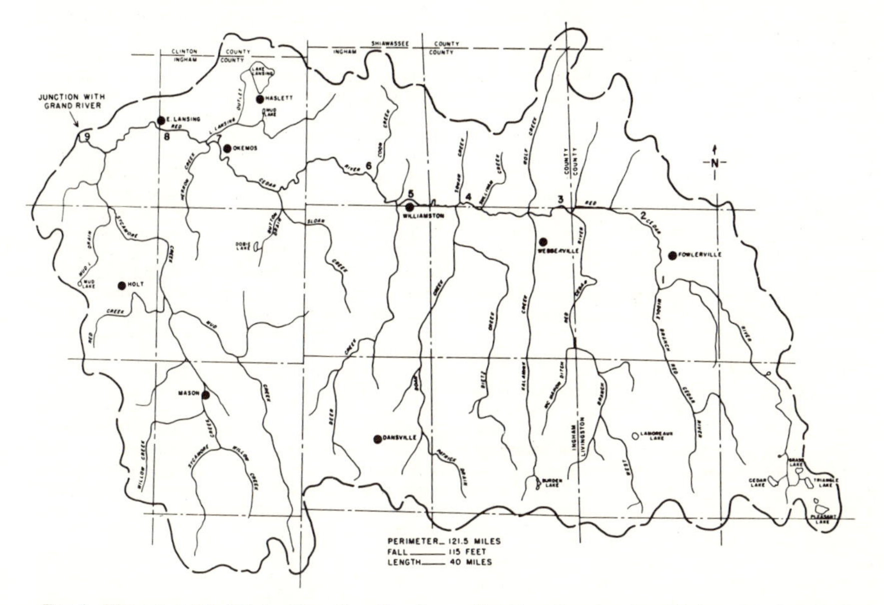

FIG. 1. *Watershed of Red Cedar River. Sampling sites used in this study are numbered 1 through 9. Compiled by R. Horner.*

bridge, the river flows into a reservoir that is impounded by a dam in Williamston. No sampling sites were located along this part of the stream. The next sampling site (number 5) was located immediately below the dam, upstream of the Williamston municipal sewage treatment facility. Site 6 was located at the Zimmer Road bridge about 2 miles below the sewage treatment plant.

The river bottom gradually changes from a rock and gravel bottom below Williamston to a sand bottom at Okemos. The section of the stream below Zimmer Road adjoins woodland much of the way, receives no urban effluent, and is the cleanest part of the entire stream. Sampling site 7 was located in Okemos at the bridge over the golf course. The next site, number 8, was at Farm Lane on the Michigan State University campus. Site 9, the last site, was located at Potter Park bridge in Lansing below the East Lansing sewage treatment plant. More information on the stream is available from Linton and Ball (8).

Sampling techniques. Water samples were collected with aseptic techniques. The sampling apparatus consisted of 1-liter glass-stoppered reagent bottles wired by their necks onto a wood pole. A string tied to the glass stopper was used to open the bottle during sample collection. At the time of collection, the sterile sampling apparatus was carefully unwrapped, and the bottle was inserted to a constant depth of 6 inches (15.24 cm) into the midstream portion of the river. After insertion, the string was pulled to open the bottle for collection. If the bridge was high above the water, an extension was attached to the sterilized unit before it was unwrapped. At two of the sites, 6 and 7, waders were worn into the river for collection of the samples. At site 5, the samples were collected from the shore about 20 m downstream of the spillway. After filling the bottle with about 900 ml of water, the stopper was replaced, and the sample was returned to the vehicle for immediate handling. First, 1-ml portions were aseptically transferred to extinction-dilution tubes for viable counting. Then, 200 ml was added to an Erlenmeyer flask containing 8 ml of 25% glutaraldehyde, *p*H 6.8. The remaining sample was stored on ice in the sampling bottle and used for *p*H determinations upon return to the laboratory. The water temperature was obtained by equilibrating a thermometer in a bucket of freshly collected river water.

Enumeration. Dilutions for viable counting were performed as soon as the water sample was returned to the vehicle. All viable counts were determined by the extinction-dilution technique using most-probable-number (MPN) tables. Total viable and viable prosthecate bacteria were enumerated in a dilute medium consisting of 0.005% peptone (Difco) and 0.005% yeast extract (Difco) with 10 ml/liter of a vitamin solution (10) and 20 ml/liter of a mineral salts solution (10). Portions (1 ml) of the river water were added to each of 10 tubes containing 9 ml of the medium. Each of these tubes was serially diluted until the highest dilution received 10^{-9} ml of the river water. Thus, 100 tubes of this medium were inoculated at each site. Upon returning to the laboratory, these were incubated at room temperature (22 to 25 C) for 5 to 7 days. The tubes were then examined macroscopically for turbidity, the criterion used for assessing total viable numbers. The tubes were incubated for an additional 2 weeks before wet mounts of each were examined individually under oil immersion with a Zeiss GFL phase microscope. Examination proceeded from the lowest dilution tube in a series and continued until one tube in the series was found negative for prosthecate bacteria. Indvidual types were identified on the basis of their morphology as caulobacters (*Caulobacter*, *Asticcacaulis*, and the fusiform caulobacter), *Hyphomicrobium*, *Prosthecomicrobium*, and *Ancalomicrobium*. This enumeration procedure required approximately 2 one-man work weeks (occasional checks of series examined earlier did not reveal any changes in the type of prosthecate bacteria resulting from prolonged incubation). Ten-tube MPN tables were obtained from Halvorson and Ziegler (6).

Coliform organisms were enumerated by using Lauryl Tryptose Broth (Difco) in the 5-tube presumptive test as outlined in *Standard Methods for the Examination of Water and Wastewater* (1). As with the other viable enumerations, these were inoculated in the field immediately after collection of the water sample and stored in the vehicle until returning to the laboratory at the end of the sampling period, which usually began at about 9:15 AM and was completed by 3:00 PM.

On one occasion, the 3 June 1968 sampling, the presumptive test was followed by a fecal coliform test to determine the proportion of fecal coliforms to total coliforms along the watercourse. The boric acid-lactose broth test (1) was used for this purpose.

Direct enumeration. The 200-ml samples that were fixed with glutaraldehyde immediately upon collection were used for the direct enumeration studies. Upon return to the laboratory, these samples were stored at refrigerator temperature until they could be centrifuged to concentrate the cells. Centrifugation (quantitative) was at 20,000 × *g* for 30 min in a Sorvall RC2-B centrifuge at 0 to 5 C. The supernatant fluid was carefully decanted, and the particulate fraction was resuspended to a known volume, usually 1.0 ml, with water and glutaraldehyde from the same bottle. These were then stored until wet mounts could be examined in the phase microscope. The proportion of each type of prosthecate bacterium was determined by examining 500 organisms from each sample. Because of the uncertainty associated with identification, the prosthecate bacteria were grouped as hyphomicrobia (*Hyphomicrobium*, *Pedomicrobium*, and *Rhodomicrobium*), prosthecomicrobia (*Prosthecomicrobium* and *Prosthecochloris*), *Ancalomicrobium*, and caulobacters (*Caulobacter* and *Asticcacaulis*). In one instance, a quantitative enumeration was made with a Petroff-Hauser counting chamber.

RESULTS

Figure 2 shows the viable count data for 7 July 1968. Total viable numbers remained relatively constant along the watercourse at about

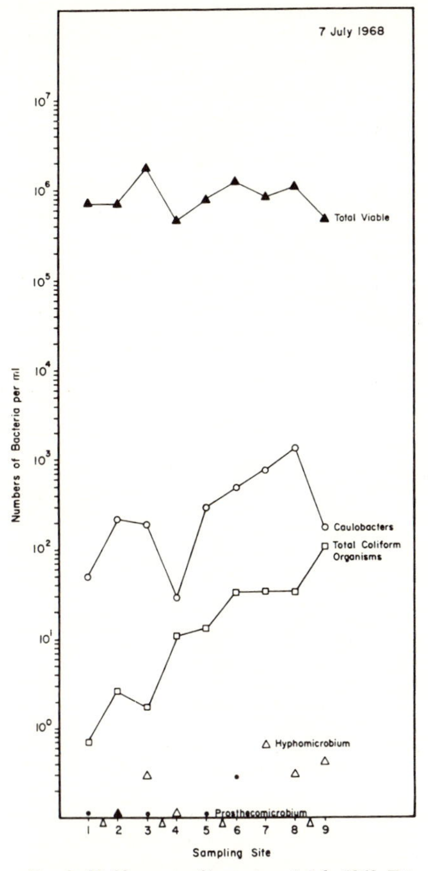

FIG. 2. *Viable counts of bacteria on 7 July 1968. Triangles between sampling sites designate location of municipal sewage treatment plant effluent discharge.*

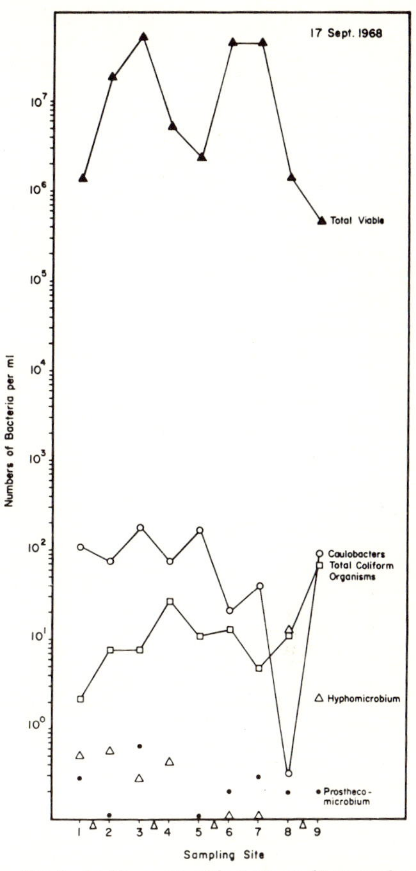

FIG. 3. *Viable counts of bacteria on 17 September 1968.*

10^6 per ml. The caulobacters varied widely from a low of 27.8 to 1,330 per ml. Total coliform organisms were fewer than the caulobacters at each site, ranging from a low of less than 1.0 per ml at site 1 to a high of about 100 per ml at site 9. *Hyphomicrobium* and *Prosthecomicrobium* were found at most sites, but their numbers (< 1.0 per ml) were not high enough to permit accurate counting. For this reason, these points on the graph were not interconnected. The water temperature during this time varied from 21.4 to 23.0 C. On the basis of direct counts, the prosthecate bacteria accounted for an average of 0.75% of the total microorganisms counted, of which 80% were caulobacters, 15% were hyphomicrobia, and 5% were prosthecomicrobia. The predominant morphological group of bacteria were the vibrios which accounted for about 60% of the total microorganisms present at those sites where they were counted.

Samples were collected again on 17 September 1968 when temperatures were 18.7 and 21.0 C. Total viable numbers (Fig. 3) in some cases were in excess of 10^7 per ml, reflecting the tremendous densities in the stream, the water of which was visibly turbid in the sampling bottles at the time of collection. The caulobacters were somewhat fewer than during the previous sampling period,

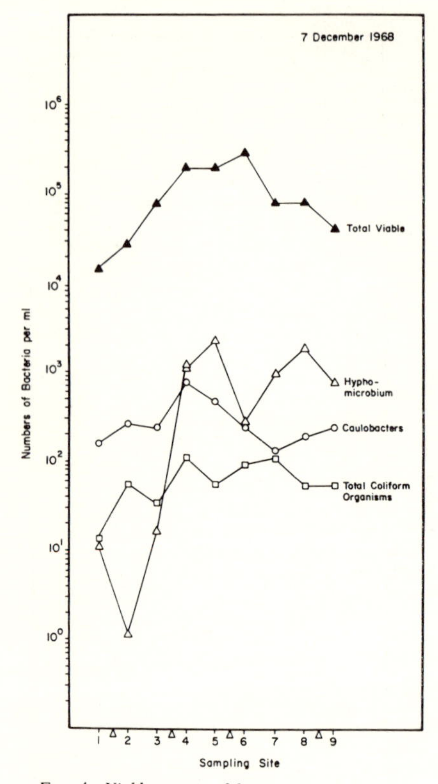

FIG. 4. *Viable counts of bacteria on 7 December 1968.*

FIG. 5. *Viable counts of bacteria on 18 March 1969.*

varying from about 20 to 188 per ml, the exception being an extremely low count at site 8. Total coliform counts were comparable to those obtained during July. *Hyphomicrobium* and *Prosthecomicrobium* were also found, but again their numbers were generally below 1 per ml. Direct counts revealed that prosthecate bacteria accounted for 0.62% of the total microflora composed of 64% caulobacters, 22% prosthecomicrobia, and 14% hyphomicrobia.

On 7 December 1968, the water temperature reached 0.0 C at several of the upstream sampling sites that were iced over or contained ice on the edges. The high temperature was recorded as 1.0 C at the last sampling site. Total viable bacteria were much lower than previous dates, ranging from 1.5×10^4 to 2.7×10^5 per ml (Fig. 4). The numbers of caulobacters, however, increased from the previous period, varying from a low of 133 to a high of 792 per ml. They were surpassed in abundance at the downstream sites by hyphomicrobia, which were the most numerous of the prosthecate bacteria at that time, reaching a high of 2,420 per ml at site 5. The coliform organisms ranged from 13 per ml at site 1 to a high of 109. Prosthecate bacteria accounted for 0.76% of the direct count. Caulobacters, dominated by members of the genus *Asticcacaulis*, were the predominant forms, amounting to 79% of the prosthecate bacteria, whereas hyphomicrobia and prosthecomicrobia constituted 15 and 6%, respectively. Vibrios were again significant members of the bacterial community accounting for 13% of the total microflora.

Figure 5 shows the data for 18 March 1969. The temperature of the water varied from a low of 6.0 C to a high of 8.0 C. The total viable count ranged from a low of 8.72×10^4 to 4.93×10^5

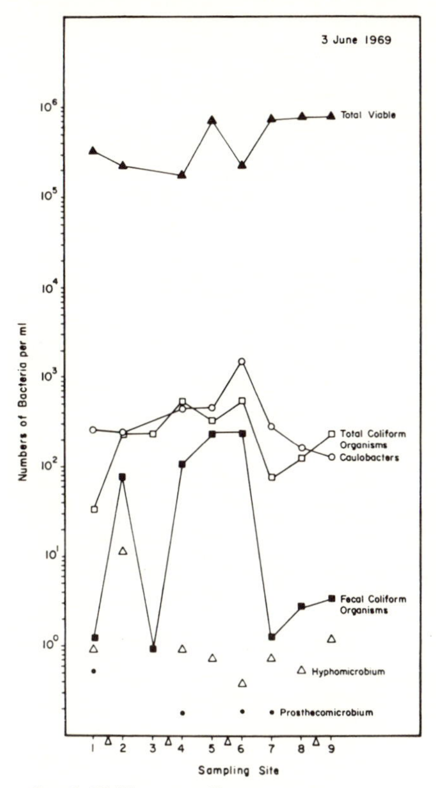

FIG. 6. *Viable counts of bacteria on 3 June 1969.*

per ml. Caulobacters reached their highest numbers recorded (3,990 per ml at site 2) and generally declined in numbers at subsequent sites to a low of 560 per ml at site 9. Coliform organisms were 70 per ml or fewer. Other prosthecate bacteria were low in numbers (< 1.0 per ml). Prosthecate bacteria were 0.93% of the total microbial flora. Again caulobacters dominated this group, consisting of 93% of the appendaged forms compared to 8% for prosthecomicrobia and 2% for *Ancalomicrobium*.

A direct quantitative count was made on the site 2 sample. By this procedure, there were 3.9×10^5 cells per ml. The viable count was 38% of this or 1.5×10^5 cells per ml.

The final river sampling was conducted on 3 June 1969 (Fig. 6). The temperature ranged from 14.0 C at the upstream sites to 16.0 C at site 9. Total viable numbers approached 10^6 per ml at several of the downstream sites. Viable caulobacters ranged from 130 per ml to 1,530 per ml, significantly lower than on the March sampling but comparable to that of the previous July. Total coliform organisms varied from a low of 35 per ml at the first site to a high of 542 at the site below the Williamston treatment plant. Fecal coliforms were also enumerated. Their numbers were lowest at site 1, being about 9 per ml, to highs of 240 per ml at sites 5 and 6 below Williamston. Again *Hyphomicrobium* and *Prosthecomicrobium* were present in detectable but low numbers. Direct counts indicated that prosthecate bacteria were 1.1% of the total microflora. Most of these were caulobacter (70%), with 24% hyphomicrobia.

DISCUSSION

This study shows that prosthecate bacteria occur in some freshwaters in large enough numbers to permit enumerating by accepted techniques for both viable and direct counting. Furthermore, the results of this study indicate that these bacteria constitute a significant portion of the microorganisms in this stream. By direct count the prosthecate bacteria comprised about 1% (range 0.62 to 1.1%) of the total microbial flora. These estimates are no doubt conservative because many of the prosthecate bacteria have nonappendaged stages in their life cycles, making them indistinguishable from other unicellular heterotrophs. This problem is especially pertinent to the caulobacters which were the most numerous of these forms (64 to 93% of total prosthecate bacteria). If one assumes that the stalked and nonstalked cells were equal in numbers and evenly distributed in the stream, then the caulobacters would have been underestimated by a factor of two with the direct-count method.

Viable enumerations also indicated that the caulobacters were usually the most numerous of the prosthecate bacteria. They ranged from fewer than 1 per ml to as many as about 4,000 per ml. Their proportion of the total viable count varied from a low at one site in September of 0.00005% to a high at one site in March of 2.8% of the total viable microflora. During one sampling period (December), *Hyphomicrobium* was more abundant than the caulobacters at the downstream sites.

The viable counts on caulobacters reported in this study compare favorably with those reported by Belyaev (3), who found from 10 to 10,000 caulobacters per ml in a variety of fresh water reservoirs along the Volga-Don rivers.

The discrepancy between the direct and viable counts during the September sampling is difficult to explain. In that month, the caulobacters comprised less than 0.002% of the total viable count at all sites, yet the direct counts indicated that they accounted for about 0.4% of the total microflora, a difference greater than two orders of magnitude. This same anomaly was observed during the other summer months, although it was not as pronounced. One explanation is that most of the caulobacters were nonviable at the time of collection. Another possible explanation is that the medium employed did not permit the growth of the most numerous caulobacters. An additional factor was that only 4,500 cells were counted for direct enumeration during each sampling period.

A correlation graph was constructed by plotting the log of total coliforms against the log of total caulobacters at each site for each date. There was no indication of a relationship between their frequencies for the study, although during individual sampling periods both negative and positive correlations were occasionally evident.

The study shows that the number of viable prosthecate bacteria varies greatly during the year. Thus caulobacters were found in high numbers on 18 March, and *Hyphomicrobium* was found in high numbers on 7 December. Although it is possible that temperature was responsible for this seasonal variation, there is not enough evidence to substantiate this, particularly in view of the large number of other variables that could conceivably influence the incidence of these bacteria.

The techniques described in this paper should permit a more thorough evaluation of the incidence of prosthecate bacteria in freshwater and hopefully lead to the determination of the factors that influence the occurrence of these organisms.

ACKNOWLEDGMENTS

This research was supported by a grant from the Institute of Water Research, Michigan State University (Annual Allotment grant 14-0-0001-1842).

I am most appreciative of the technical assistance provided by Lois Stiffler, J. Carter, J. A. M. de Bont, and C. S. Chopra.

LITERATURE CITED

1. American Public Health Association, 1967. Standard methods for the examination of water and wastewater, 12th ed. New York.
2. Aristovskaya, T. V. 1961. Accumulation of iron in breakdown of organomineral humus complexes by microorganisms. (English translation). Dokl. Akad. Nauk. SSSR **136**:954–957.
3. Belyaev, S. S. 1967. Distribution of the caulobacter group of bacteria. (English translation) Mikrobiologiya **36**:157–162.
4. Duchow, E., and H. C. Douglas. 1949. *Rhodomicrobium vannielii,* a new photoheterotrophic bacterium. J. Bacteriol. **58**:409–416.
5. Gorlenko, V. M. 1970. A new phototrophic green sulphur bacterium—*Prosthecochloris aestaurii nov. gen. nov. spec.* Z. Allgem. Mikrobiol. **10**:147–149.
6. Halvorson, H. O., and N. R. Ziegler. 1933. Application of statistics to problems in bacteriology. I. A means of determining bacterial population by the dilution method. J. Bacteriol. **25**:101–121.
7. Hirsch, P., and S. R. Conti. 1964. Biology of budding bacteria. I. Enrichment, isolation and morphology of *Hyphomicrobium spp.* Arch. Mikrobiol. **42**:17–35.
8. Linton, K. J., and R. C. Ball. 1965. A study of the fish populations in a warm-water stream. Quart. Bull. Mich. Agr. Exp. Station **48**:255–285.
9. Poindexter, J. S. 1964. Biological properties and classification of the *Caulobacter* group. Bacteriol. Rev. **28**:231–295.
10. Staley, J. T. 1969. *Prosthecomicrobium* and *Ancalomicrobium*: new prosthecate freshwater bacteria. J. Bacteriol. **95**:1921–1942.

NATURAL RELATIONSHIPS OF INDICATOR AND PATHOGENIC BACTERIA IN STREAM WATERS

R. Jay Smith and Robert M. Twedt

Water quality is routinely evaluated by determining the concentration of coliform bacteria.[1] The presence of fecal coliforms (FC) and fecal streptococci (FS) results from recent contamination by human or animal excrement. According to Geldreich and Kenner,[2] the FC:FS ratio may indicate the probable source, and, therefore, clarify the significance of pollution. They found that FC:FS ratios were always greater than 4.0 in the feces of man and in domestic wastewater, while they were less than 0.7 in the feces of farm animals, cats, dogs, and rodents and in wastewater polluted with such feces. Some recent water quality surveys that deal quantitatively with indicator concentrations have described limited qualitative pathogen recoveries.[3,4] At the present time, however, there is little information available on the quantitative relationships between indicator organisms, such as the coliforms and streptococci, and pathogens such as *Salmonella*. Such information is basic to the bacteriological indicator concept. It was the purpose of this study to conduct an extensive quantitative bacteriological examination of tributary streams of varying quality. From the stream profiles thus derived, it was proposed to study the natural relationships of indicator organisms and *Salmonella*.

Materials and Methods

An average of 13 water samples were collected at each sampling site from June to October 1968. The pH, stream temperature, and dissolved oxygen (DO) of the river water were determined at the time and place of sampling with a portable pH meter,* and portable oxygen analyzer † which was calibrated weekly by the Winkler method. Total coliforms (TC) were determined according to membrane filtration (MF) procedures outlined in "Standard Methods."[1] Fecal coliform determinations were carried out and verified according to the MF technique described by Geldreich *et al.*[5] FS were quantitated by an MF technique described by Kenner *et al.*[6] At appropriate intervals, MF procedures were validated by comparison with multiple tube dilution, most probable number (MPN) techniques.[1]

Salmonellae were determined by an MPN technique according to Raj.[7] One 50-ml, five 10-ml, and five 1-ml portions of each sample were inoculated into 50 ml of double-strength, 10 ml of double-strength, and 10 ml of single-strength dulcitol selenite enrichment, respectively. Tubes were incubated at 41.5°C and checked for selenite reduction after 24 and 48 hr. Positive tubes were streaked on MacConkey's agar and either bismuth sulfite agar, brilliant green agar, or

* *Model 175, Instrumentation Laboratories, Inc., Boston, Mass.*

† *Model 51, Yellow Springs Instrument Co., Inc.. Yellow Springs, Ohio.*

xylose lysine brillant green agar.[8] The plates were examined for possible *Salmonella* and all such colonies were transferred to triple sugar iron agar, urea agar, SIM medium, lactose broth, and dulcitol broth. Suspected salmonellae were screened with polyvalent 0 antisera provided by the Michigan Department of Public Health and with polyvalent H antisera. Final serotyping was carried out at the Communicable Disease Center, Atlanta, Ga.

Descriptions of the River Systems Studied

Ten representative collecting sites were regularly sampled along the Saline River (Table I). The Saline River (Figure 1), located in Washtenaw and Monroe counties in southeastern Michigan, courses in a southeasterly direction through agricultural and undeveloped rural, suburban, and industrial areas. The Saline, approxi-

TABLE I.—Saline River Sampling Sites

Station Number in Miles from Source	Location
SR 3	Brumeister and Lima Center Roads
SR 10	Austin and Schill Roads
SR 11	Dell Road
SR 13	Impoundment at Highway U. S. 12
SR 14	Saline Municipal Park, upstream from die-casting company effluents
SR 15	Macon Road, between WWTP and die-casting company effluents
SR 16	One mile south of Saline city limits
SR 18	Milky and Maple Roads
SR 23	Dennison Road
SR 33	Allison Road

Note: Miles × 1.6 = km.

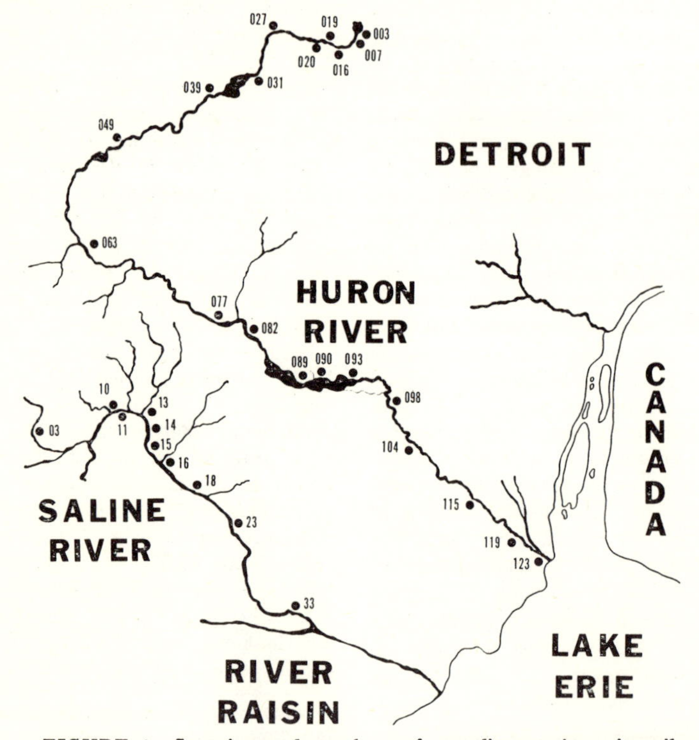

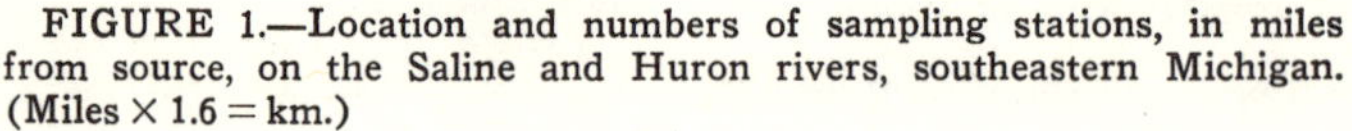
FIGURE 1.—Location and numbers of sampling stations, in miles from source, on the Saline and Huron rivers, southeastern Michigan. (Miles × 1.6 = km.)

TABLE II.—Huron River Sampling Sites

Station Number in Miles from Source	Location
Upper Huron River*	
HR 3	White Lake Road at Teggerdine Road
HR 7	Highland Road
HR 16	Cooley Lake Road
HR 19	Commerce Road
HR 20	At Lower Straits Lake outlet
HR 27	River Road in Milford
HR 31	Dawson Road at inlet of Kent Lake
HR 39	Kenningston Road at outlet of Kent Lake
HR 49	At U. S. 36 highway overpass
HR 63	Upper Huron Metro Park
HR 77	Fuller Road in Ann Arbor
Lower Huron River*	
HR 82	Superior Avenue in Ypsilanti
HR 89	Bridge Street at outlet of Ford Lake
HR 90	Belleville Lake at outlet of Willow Creek
HR 93	Denton Road Bridge at outlet of Belleville Lake
HR 98	Lower Huron Metro Park
HR 104	Waltz Street Bridge in New Boston
HR 115	Two miles upstream from Telegraph Road in Flat Rock
HR 119	Fort Street in South Rockwood
HR 123	Point Mouillee State Game Refuge

* Upper and lower Huron are arbitrary designations to facilitate analysis of data.
Note: Miles × 1.6 = km.

mately 35 miles (56 km) in length, is a tributary of the River Raisin which, in turn, discharges into the basin of western Lake Erie. The mean monthly discharge of the Saline River, recorded at U. S. Geological Survey (USGS) gauging station 4–1764 near the city of Saline, Mich., was 262, 124, 66.9, 30.6, and 23.5 cfs (445, 210, 114, 52, and 40 cu m/min) from June to October 1968, respectively.[9] During the period sampled, the Saline River basin supported a total population of approximately 20,000 people, most of whom were located in the small communities of Bridgewater, Saline, Mooreville, and Milan. Among other industrial complexes in the immediate area of Saline are a die-casting company and a ball and bearing manufacturing company. Saline is the largest and most heavily industrialized of these communities. During the period sampled, the die-casting regularly discharged toxic effluents containing cyanide, trivalent chromium, nickel, and copper into the river. The spectrographic analyses showed trace amounts of chromium and cyanide in the river a short distance above the plant. Here zinc concentration was 0.1 mg/l in one of the analyses, and nickel concentration was 0.005 mg/l. Immediately downstream from the die-casting company, the nickel concentration rose to 0.285 mg/l. Four out of nine such sampling stations showed concentrations of nickel above 0.07 mg/l, all of these being below the die-casting company. A biological-chemical survey of the Saline River was undertaken by the Michigan Water Resources Commission during June 1968.[10] Their purpose was to relate Saline River water quality and benthic fauna with wastewater treatment plant and manufacturing effluent discharges. They found that the Saline treatment plant and the die-casting effluents exerted considerable influence on the character of the water downstream, particularly in the amounts of total chromium, nickel, soluble orthophosphates, and ammonia being introduced via these effluents into the river. This report indicated total chromium and nickel in concentrations of 0.04 mg/l and 0.5 mg/l, respectively, slightly downstream from the effluent source. Concentrations slightly under these amounts were still detected distances of 12 miles (19 km) downstream and farther.

TABLE III.—Average Concentrations of Indicator and *Salmonella* Organisms in the Saline River, June through October 1968

Site	Total Coliforms (no./100 ml)	Fecal Coliforms (no./100 ml)	Fecal Streptococci (no./100 ml)	Ratio FC:FS	Salmonellae (no./100 ml*)	Ratio TC:S	Ratio FC:S	Ratio FS:S
SR 3	24,000	3,200	3,200	1.0	1	60,000	4,733	20,000
SR 10	16,000	1,300	3,200	0.4	—	—	—	—
SR 11	15,000	1,200	3,000	0.4	—	—	—	—
SR 13	5,800	920	2,100	0.4	—	—	—	—
SR 14	11,000	1,100	2,900	0.4	1	22,333	2,650	1,445
SR 15	13.000	1,100	1,200	0.9	—	—	—	—
SR 16	17,000	1,000	1,700	0.5	—	—	—	—
SR 18	21,000	1,900	1,900	1.0	—	—	—	—
SR 23	21,000	2,400	2,200	1.0	1	5,533	630	124
SR 33	56,000	3,500	3,800	0.9	1	159,000	7,100	160
						32,950†	2,737†	8,702†

* *Salmonella* (S) concentrations are averages of samples from which pathogens were isolated.
† Geometric means.

Twenty representative collecting sites were sampled regularly along the Huron River (Figure 1 and Table II). Eleven sites were in the upper part and nine were in the lower part of the river. The Huron River drains an area of approximately 892 sq miles (2,280 sq km) in southeastern Michigan, and it meanders for nearly 123 miles (197 km), including all of its river lakes. In its course, it passes through mixed rural, urban, and many highly industrialized areas, from its source in the lakes region of Oakland County north of Detroit, Mich., to its discharge into western Lake Erie. The mouth of the Huron River widens into an extensive marshland area, the Point Mouilee game refuge. This general area also includes the mouth of the Detroit River where it empties into Lake Erie. The mean monthly discharge of the Huron River, recorded at USGS Gauging Station 4.1745 at Ann Arbor, Mich., was 1,148, 1,130, 569, 377, and 332 cfs (1,960, 1,920, 970, 640, and 565 cu m/min) from June to October, respectively.[9] The Huron River basin, during the period studied, supported a total population of approximately 246,000 people, of whom about 127,000 resided in and around the adjacent cities of Ann Arbor and Ypsilanti, Mich. The remainder of the population could be accounted for in the small suburban communities north and southeast of Detroit, Mich.

Results

Saline River

The average concentration of indicator organisms and salmonellae found at the 10 Saline River sites during 1968 are shown in Table III. The TC, FC, and FS concentrations were 920 organisms/100 ml or greater at all sites, but no significant fluctuations could be noted. FC comprised less than 16 percent of the TC concentrations, the greatest percentage occurring in the impounded river water in Saline.

At 5 of the 10 sites, an average FC:FS ratio of 0.4 strongly suggested that animal waste was the principal source of pollution. At the remaining 5 of the 10 sites, FC:FS ratios of approximately 1.0 indicated a mixture of human and animal waste pollutants.

Salmonellae were isolated six times from four sites on the Saline River. Three of the four had average FC:FS ratios of approximately 1.0. *S. typhimurium* var. *copenhagen* was isolated at Saline River (SR) 3 and SR 14, *S. typhimurium* at SR 33, *S. enteritidis* at SR 23, and *S. java* at SR 14 and SR 23. In the samples from which

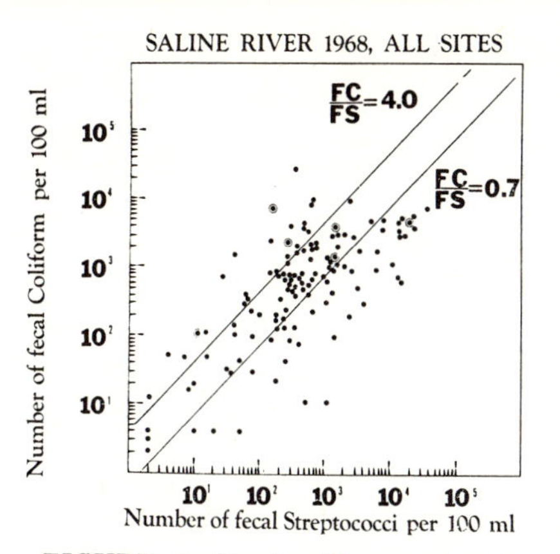

FIGURE 2.—Fecal coliform concentrations compared to fecal streptococci concentrations and the occurrence of salmonellae in Saline River, June through October 1968. Circled points indicate isolation of salmonellae.

salmonellae were isolated, the ratios of geometric means relating indicators to pathogens were 32,960 coliforms, 2,737 FC, and 8,702 FS/salmonellae.

In Figure 2, the FC concentration is plotted against the FS concentration of each sample taken from the Saline River. A portion of the 126 samples had FC:FS ratios greater than 4.0 (24 samples) or less than 0.7 (42 samples), the limits suggested by Geldreich and Kenner [2] as characteristic of pollution by human or animal waste, respectively. However, many of the samples (60) exhibited ratios between these extremes, characteristic of mixed pollution. Three of the six *Salmonella* isolations occurred when FC:FS ratios were greater than 4.0 and one when the ratio was less than 0.7. No salmonellae were isolated when the FC concentration was less than 100 organisms/100 ml.

Huron River

The average indicator bacteria concentrations for 11 sampling sites on the upper Huron River during 1968 appear in Table IV. The average total coliform concentration at the source (5,700 organisms/100 ml), decreased to 950 organisms/100 ml at the first station below Kent Lake, Huron River (HR) 39. The levels rose steadily to a high of 14,000 organisms/100 ml at HR 77 in Ann Arbor. The FC concentrations followed a similar pattern. From an average of 500 organisms/100 ml at the source, the concentration fell to 46 organisms/100 ml immediately below Kent Lake. At HR 77 in Ann Arbor, the FC level increased to 500 organisms/100 ml. The FC in the upper Huron River comprised less than 10 percent of the total coliform population. The smallest fraction, found at HR 77, was 2.4 percent. The concentrations of FS decreased from 1,000 organisms/100 ml at HR 3 to 29 organisms/100 ml at HR 39 and then increased to 740 organisms/100 ml at HR 77. Of the 11 sites, 7 exhibited FC:FS ratios less than 0.7, characteristic of animal source pollution. Four ratios were greater than 1.0, indicating a mixture of human and animal waste. When the FC concentrations were plotted against the FS concentrations for all 91 samples from the upper Huron River (Figure 3), most samples (55) exhibited FC:FS ratios between 4.0 and 0.7. However, fully a third (31) exhibited ratios less than 0.7, while only five were greater than 4.0

TABLE IV.—Average Concentrations of Indicator Organisms in the Upper Huron River, June through October 1968

Site	Total Coliform (no./100 ml)	Fecal Coliform (no./100 ml)	Fecal Streptococci (no./100 ml)	Ratio FC:FS
HR 3	5,700	500	1,000	0.5
HR 7	5,900	450	720	0.6
HR 16	3,700	220	740	0.3
HR 19	2,300	80	120	0.7
HR 20	3,500	180	490	0.4
HR 27	2,100	110	250	0.4
HR 31	1,400	83	67	1.2
HR 39	950	46	29	1.6
HR 49	3,100	320	130	2.5
HR 63	6,100	180	170	1.0
HR 77	14,000	340	740	0.4

Salmonellae were not isolated from the upper Huron River from June through October 1968.

Coliform densities in the lower Huron River, as seen in Table V, were similar to those in the upper Huron during the 1968 study. Average TC and FC levels ranged from 5,200 to 12,000 organisms/100 ml and from 86 to 820 organisms/100 ml, respectively. FC varied from 1 to 7.5 percent of the TC. The FS concentrations seemed to differ materially in the lower and upper Huron River. In the lower portion, they ranged from 10 to 1,500 organisms/100 ml, a decline from the upper Huron averages of 30 to 7,300 organisms/100 ml. Consequently, at five of the sampling sites on the lower Huron, average FC: FS ratios were greater than 1.0. Salmonellae were isolated in the lower Huron at HR 82 (*S. kottbus*) and HR 119 (*S. typhimurium* var. *copenhagen*). In the samples from which salmonellae were isolated, the ratios of geometric means relating indicators to pathogens were 11,580 coliforms, 300 FC, and 191 FS/salmonellae. One-half (57) of all 105 samples had FC: FS ratios between 4.0 and 0.7 when FC concentrations from the lower Huron River were plotted against the FS concentrations (Figure 4). Approximately one-fourth (22) exhibited ratios less than 0.7, while another one-fourth (26) were greater than 4.0. Salmonellae were isolated from the lower Huron River from two samples having FC:FS ratios greater than 0.7 but less than 4.0. No salmonellae were isolated when the FC concentration was less than 200 organisms/100 ml.

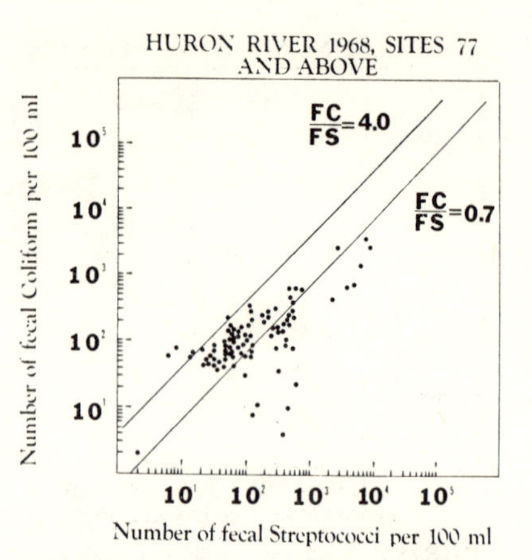

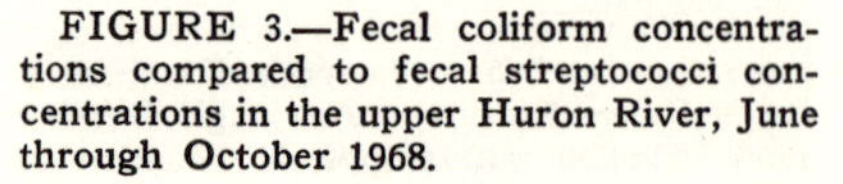

FIGURE 3.—Fecal coliform concentrations compared to fecal streptococci concentrations in the upper Huron River, June through October 1968.

Discussion

It is almost routine for sanitary stream surveys to present quantitative TC, FC, and FS data together with supportive qualitative isolations of *Salmonella*. Quite often the latter may be associated with sampling sites

TABLE V.—Average Concentrations of Indicator and *Salmonella* Organisms in the Lower Huron River, June through October 1968

Site	Total Coliforms (no./100 ml)	Fecal Coliforms (no./100 ml)	Fecal Streptococci (no./100 ml.)	Ratio FC:FS	Salmonellae (no./100 ml)*	Ratio TC:S	Ratio FC:S	Ratio FS:S
HR 82	12,000	610	270	2.2	2	31,667	375	175
HR 89	5,200	130	86	1.5	—	—	—	—
HR 90	5,400	86	53	1.6	—	—	—	—
HR 93	7,400	93	12	7.8	—	—	—	—
HR 98	6,200	110	120	0.9	—	—	—	—
HR 104	6,400	100	250	0.4	—	—	—	—
HR 115	8,100	210	320	0.6	—	—	—	—
HR 119	11,000	710	1,000	0.7	1	4,333	240	210
HR 123	11,000	820	1,500	0.5	—	11,580†	300†	191†

* *Salmonella* (S) concentrations are averages of samples from which pathogens were isolated.
† Geometric means.

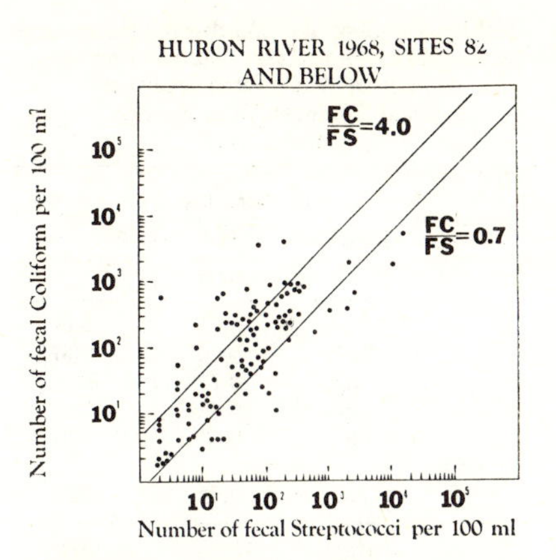

FIGURE 4.—Fecal coliform concentrations compared to fecal streptococci concentrations and occurrence of salmonellae in lower Huron River, June through October 1968. Circled points indicate isolation of salmonellae.

or river distance from the pollutional source.[3, 4] Unfortunately, few if any data are available that quantitatively relate indicator and *Salmonella* concentrations.

In a recent review, Geldreich [11] emphasized that use of a given bacteriological parameter of water quality "must be ultimately related to the probable occurrence of waterborne pathogens." Summarizing data gathered from numerous streams and estuarine pollution studies, he demonstrated that *Salmonella* isolation occurred in less than 27.6 percent of freshwater samples when the FC concentration was less than 200 organisms/100 ml. The frequency of isolation rose sharply to 85.2 percent when the range of FC was 200 to 2,000 organisms/100 ml and 98.1 percent when there were more than 20,000/100 ml. In a qualified projection from these data, Geldreich suggested that the 200 FC/100 ml limiting concentration "may be of useful water quality value."

Van Donsel and Geldreich [12] examined the relationships of *Salmonella* and FC in bottom muds taken from freshwater lakes and streams of varying size and quality. They recovered salmonellae from two-thirds of mud samples when the water above contained more than 200 FC/100 ml but only infrequently when the concentration was less than 200 organisms/100 ml. The authors commented that the level of 200 FC/100 ml might represent a significant limiting relationship between indicator and pathogen or merely result from limitations in present *Salmonella* methodology. Their results were strikingly similar, however, to the qualitative isolation data presented by Geldreich.[11]

The data of this study support the suggested limiting FC value of 200 organisms/100 ml using quantitative MPN methods. In this study salmonellae have not been isolated from the Saline River when the FC were less than 100 organisms/100 ml, nor from the Huron River when the concentration was less than 200 organisms/100 ml.

Salmonella isolations are less frequent with the quantitative MPN technique used here than by concentration in a gauze pad or bottom sediment. The quantitative relationships between indicators and pathogens were not as well clarified by examining a natural stream profile of average bacterial concentrations as they were by the scatter plot diagrams. From the natural relationships of the former, ratios of 32,960 coliforms, 2,737 FC, and 8,702 FS in the Saline River; and 11,580 coliforms, 300 FC, and 191 FS/salmonellae in the Huron River were determined from geometric means.

The only quantitative ratios presently available were reported by Van Donsel and Geldreich [12] and Dunlop *et al.*[13] The former related one salmonellae in mud to 14,000 FC in mud and 150 FC in the overlying water. The latter related one salmonella organism to 255,000 coliforms and 4,800 enterococci in irrigation water. These figures may be corrected to extrapolate the FC concentration in irrigation water samples polluted almost entirely with fresh primary wastewater treatment effluent and to allow for 4.5 percent recovery of salmonellae by the techniques used. The corrected figure is one salmonellae for 870 fecal coliforms.

Both the Huron and Saline rivers drain basins of relatively large *Salmonella* populations. Consequently, the presence of excreters of salmonellae, either humans discharging waste directly into streams or animals discharging via stormwater runoff, seems assured. The minimum human population necessary to contribute salmonellae has been reported at less than 1 percent in sewered populations in the U. S.[14] and Great Britain.[15] Similarly, infection rates in animal populations range from 13 percent in cattle and 15 percent in sheep to 22 percent in pigs.[16] Therefore, reaches of either river flowing through suburban areas of relatively low human population density might be expected to yield samples with no *Salmonella* isolation but high FC: FS ratios. Conversely, samples from rural areas might yield salmonellae but exhibit low FC: FS ratios.

The Directory of Wastewater Treatment Works for the State of Michigan as of February 1968 lists a total of 13 wastewater treatment plants discharging effluents into the Huron River and two into the Saline River. The former directly served an estimated combined population of 127,000 people, whereas the latter directly served a population of about 6,000. As is indicated in Table VI, certain collecting sites were directly downstream from the effluent outfall of municipal wastewater treatment plants in Milford (HR 27) (trickling filter); Ann Arbor (HR 82) (activated sludge); Rockwood (HR 119) (primary treatment); Flat Rock (HR 115) (primary treatment), and Saline (SR 15) (trickling filter).

TABLE VI.—Average Discharge Rates and Post-Chlorination Total Coliform Concentrations in Wastewater Treatment Plant Effluents, June through October 1968

Treatment Plant Location	Month	Mean Discharge of Plant Effluent (mgd)	Total Coliforms (no./100 ml)
Milford	June	0.65	565
	July	0.70	500
	Aug.	0.77	211
	Sept.	0.61	745
	Oct.	0.51	1,630
	Avg.	0.64	730
Ann Arbor	June	13.80	338
	July	15.00	155
	Aug.	14.60	2,610
	Sept.	14.20	2,650
	Oct.	13.50	1,060
	Avg.	14.20	1,362
Flat Rock	June	0.72	63
	July	0.79	306
	Aug.	0.54	102
	Sept.	0.46	121
	Oct.	0.37	34
	Avg.	0.65	125
Rockwood	June	0.38	28
	July	0.44	86
	Aug.	0.46	127
	Sept.	0.18	374
	Oct.	0.16	91
	Avg.	0.32	141
Saline	June	1.00	888
	July	1.20	750
	Aug.	1.50	1,832
	Sept.	1.00	1,500
	Oct.	—	—
	Avg.	1.30	1,242

Note: Mgd $\times$ 3.785 $\times$ 10^{-3} = cu m/day.

As can be seen in Table VI, the effluent from the Milford treatment plant upriver from HR 31 showed a 5-month (June through October) average flow rate of 0.64 mgd (2,400 cu m/day) with an average of 730 TC organisms/100 ml. Similar figures were 14.2 (54,000) and 1,362; 0.65 (2,400) and 125; 0.32 (1,200) and 141, and 1.3 (4,900) and 1,242 for treatment plants at Ann Arbor, Flat Rock, Rockwood, and Saline, respectively. The mean TC concentrations for the same 5-month period at the nearest downriver stations from each of these effluents were 1,400 organisms/100 ml at HR 31; 12,000 at HR 82; 11,000 at HR 112; 11,000 at HR 123; and 17,000 at SR 16. Only slight significance could be attributed to correlations between TC concentrations in plant effluents and in the naturally flowing stream water. This suggests that, with the exception of the Milford plant, upstream of HR 31, the effluent volumes represent only a portion of the total river volume and consequently deliver only a fraction of the TC concentration.

In a correlation matrix from a computer analysis of the lower Huron River, it was revealed that the most positive correlations among the various parameters were TC to FC to flow rate, FC to FS to rainfall to flow rate, FS to flow rate, and rainfall to flow rate. Parameters that seemed to be in any way negatively correlated were FC to pH, FS to pH, DO to temperature, and rainfall to temperature. A computer analysis was also made from data accumulated from the 5-month study of the Saline River. Positive correlations were TC to FC, FC to flow rate, and rainfall to flow rate.

Studies currently being conducted in this laboratory to clarify indicator-pathogen relationships further include an intensive seasonal bacteriological examination of the two streams at selected sampling sites shown in the present work to yield *Salmonella* and correlated studies of survival in dialysis bags suspended in the streams at certain of these sites.

Acknowledgments

The authors wish to give recognition to Marianne Hahn, without whose efforts and perseverance this work would have been curtailed. Special thanks are also due Barbara Leonard and Frank Jaszcz for their excellent technical assistance and to Linda Krusel Flannigan for her assistance in the preparation of the manuscript. The authors also wish to thank the Michigan Department of Public Health for providing antisera, and the Communicable Disease Center in Atlanta, Ga., for the verification of serotyping. Recognition should also be given to Paul Bent and Lawrence Stoimenoff of the United States Geological Survey, Lansing, Mich., for their cooperation in providing flow rate data.

Special mention should be afforded Paul Blakeslee of the Michigan Department of Public Health and Edwin Geldreich of Taft Engineering Center, Cincinnati, Ohio, and Robert H. Bordner of the Environmental Protection Agency, Office of Water Programs, Cincinnati, Ohio.

This research was supported by Federal Water Quality Administration Grant No. 16030.

References

1. "Standard Methods for the Examination of Water and Wastewater." 12th Ed., Amer. Pub. Health Assn., New York, N.Y., (1965).
2. Geldreich, E. E., and Kenner, B. A., "Concepts of Fecal Streptococci in Stream Pollution." *Jour. Water Poll. Control Fed.*, **41**, R336 (1969).
3. Spino, D. F., "Elevated-Temperature Technique for the Isolation of *Salmonella* from Streams." *Appl. Microbiol.*, **14**, 591 (1966).
4. Brezenski, F. T., and Russomanno, R., "The Detection and Use of Salmonellae in Studying Polluted Tidal Estuaries." *Jour. Water Poll. Control Fed.*, **41**, 725 (1969).
5. Geldreich, E. E., *et al.*, "Fecal-Coliform Organism Medium for the Membrane Filter Technique." *Jour. Amer. Water Works Assn.*, **57**, 208 (1965).
6. Kenner, B. A., *et al.*, "Fecal Streptococci. II. Quantification of Streptococci in Feces." *Amer. Jour. Pub. Health*, **50**, 1553 (1960).
7. Raj, H., "Enrichment Medium for Selection of *Salmonella* from Fish Homogenates." *Appl. Microbiol.*, **14**, 12 (1966).

8. Taylor, W., "Isolation of Shigellae. II. New Plating Media for Isolation of Enteric Pathogens." *Bacteriol. Proc.*, **7**, 55 (1964).
9. "Water Resources Data for Michigan. Part 1. Surface Water Records." U. S. Dept. of the Int., U. S. Geol. Surv., Washington, D. C. (1968).
10. "Biological Investigations of the Saline River Vicinity of Hoover Ball and Bearing Company, Universal Die Casting Division, Saline, Michigan." Water Res. Comm., Mich. Dept. of Natl. Res. (1968).
11. Geldreich, E. E., "Applying Bacteriological Parameters to Recreational Water Quality." *Jour. Amer. Water Works Assn.*, **62**, 113 (1970).
12. Van Donsel, D. J., and Geldreich, E. E., "*Salmonella*-Fecal Coliform Relationships in Bottom Sediment." *Bacteriol. Proc.*, 18 (1970).
13. Dunlop, S. G., *et al.*, "Quantitative Estimation of *Salmonella* in Irrigation Water." *Sew. & Ind. Wastes.*, **24**, 1015 (1952).
14. Hall, H. E., and Hauser, G. H., "Examination of Feces from Food Handlers for Salmonellae, Shigellae, Enteropathogenic *Escherichia coli*, and *Clostridium perfringens.*" *Appl. Microbiol.*, **14**, 928 (1966).
15. "Excretion of *Salmonella* and *Shigella* Organisms and Enteropathogenic *Escherichia coli* in Normal Children." Report of Public Health Laboratory Service and Society of Medical Officer of Health, *Monthly Bull., Ministry Health & Pub. Health Lab. Serv., London*, **24**, 376 (1965).
16. Prost, E., and Riemann, H., "Food-borne Salmonellosis." *Ann. Rev. Microbiol.*, **21**, 495 (1967).

SALMONELLA—ITS PRESENCE IN AND REMOVAL FROM A WASTEWATER SYSTEM

E. H. Kampelmacher and Lucretia M. van Noorle Jansen

Over the last 10 yrs., an average of 4,500 cases of human salmonellosis have been recorded in the Netherlands every year. It is estimated that this figure represents only 5 to 10 percent of the true number of cases. Some patients continue to excrete salmonella in the feces for weeks or months. Detailed investigations in the Netherlands have also demonstrated that a large percentage of the livestock (e.g., 12 percent of all pigs) are clinically healthy salmonella carriers (1) (2) (3). Foods and feeds (meat, egg products, and fishmeal) also have frequently been found to be contaminated with salmonella (1) (4). As a result, both domestic wastewater and surface water in the Netherlands show a high degree of contamination with salmonella (5). The present study was made to determine the degree of contamination of wastewater by qualitative and quantitative methods in a waste treatment plant in a large Dutch town (over 100,000 inhabitants). An attempt was also made to establish whether, and if so to what extent, salmonella is eliminated by this treatment process.

Plant Description

The plant concerned currently has a capacity of 48,000 cu m/day for 325,000 population equivalents (i.e. 225,000 inhabitants and 100,000 population equivalents for industry). The three primary sedimentation basins (volume per basin, 3,300 cu m), the six high-rate trickling filters (volume per bed, 2,500 cu m) and the three secondary sedimentation basins (volume per basin 2,500 cu m) are linked in such a way that the operation of the plant may be adapted to the amount of influent. A diagram of the wastewater treatment plant is shown in Figure 1.

In the experiments to be described, the plant was circuited so that 1,500 cu m/hr of wastewater passed through one primary sedimentation basin, two parallel-circuited trickling filters, and one secondary sedimentation basin.

System Detention Time

An investigation into the residence times of water in the treatment plant was made with the aid of the radioactive nuclide Na-24. Measurements have been performed in the first and second sedimentation basin and the oxidation basin separately and in the whole installation. The activity of the Na-24 tracer put into the water was determined by a gamma scintillation counter and was registered continuously as a function of the time. Given a flow rate of 1,500 cu m/hr, the average residence time of the wastewater in the whole treatment unit was found to be 6.0 hr, while the smallest dilution volume was 9,000 cu m. This means that at the time when the maximum concentration of Na-24 was measured, the amount of Na-24 added should have dissolved homogeneously in 9,000 cu m of water. It was also established that 90 percent of the salt added had

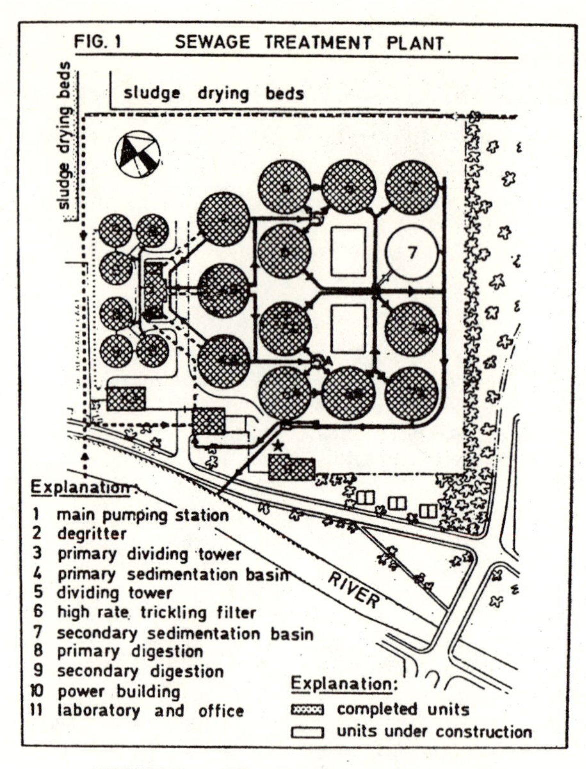

FIGURE 1.—Wastewater treatment plant.

disappeared from the plant after 10.5 hr and 99 percent after about 20 hr.

Materials and Methods

Enrichment media used in isolating salmonella from influent and effluent were Muller-Kauffmann tetrathionate broth (MK) (6) and a tenfold concentrate of selenite culture medium according to Leifson (CS) (Merck). After 18 to 20 hr and 45 to 48 hr of incubation at 37°C, smears were made on 14-cm brilliant green-phenol-red-agarplates (Br) (6). The plates were incubated at 37°C for 18 to 20 hr. Suspected colonies were transferred to triple sugar iron agar (Oxoid) and lysin decarboxylates medium (7). Serological typing followed a positive biochemical finding.

The samples of effluent to be examined for *S. utrecht* were inoculated in MK. After 18 to 20 hr, and 45 to 48 hr at 43°C (8) (9), smears were made on Br. The investigation then followed the course described above.

Examination for Salmonella

Wastewater was sampled on Wednesdays of three successive weeks in September. The influent sample was taken at the site where it enters the degritter (Figure 1, No. 2); the effluent was taken from the effluent channel (Figure 1, ★). At the first sampling of influent and effluent, a mixed sample of each was collected from 9 AM to 12 NOON; at the second and third samplings, influent samples were taken at 9 AM and effluent samples at 2 PM. Examinations for salmonella were made semiquantitatively. For this purpose, tenfold examinations were made of 0.1 ml, 1 ml, and 10 ml of influent and effluent in 9 ml, 9 ml, and 90 ml of MK and 10 ml and 100 ml of influent and effluent with 1 ml and 10 ml of CS, respectively.

Study of Elimination of Salmonellae

In an effort to establish whether, and if so to what extent, salmonellae were eliminated during treatment in the plant, a loading test was carried out. For this purpose, a 300-ml broth culture of *S. utrecht* with a germ count of 1×10^9/ml (obtained by incubation of *S. utrecht* in broth for 16 hr at 37°C) was poured into the influent in the primary dividing tower (Figure 1, No. 3) at 9 AM. *S. utrecht* was selected because this salmonella of the 0 52 group is extremely rare while others of this group are never found in the Netherlands. It could therefore be maintained with a high degree of certainty that any 0 52 Salmonella found in the effluent must have been the *S. utrecht* added to the influent.

Because study of the residence times of wastewater through the plant had shown that the smallest dilution volume was 9,000 cu m, 3×10^{11} *S. utrecht* was poured into the influent so that 3×10^3/100 ml *S. utrecht* would be present in the effluent had elimination not taken place. Because 90 percent of the Na-24 added had disappeared after 10.5 hr, it was predicted that only 10 percent of the *S. utrecht* added would be present after this interval. Three hundred salmonella germs/100 ml could still be recovered in this material. In view of the results of determinations made with Na-24, sampling was carried out for 14 hr after addition of the culture.

The samples of effluent were taken from the effluent channel (Figure 1, ★). Sampling started at 10 AM and was continued at half-hour intervals until 11 PM. A total of 27 samples of 150 ml of effluent were collected. Immediately after the sample was taken, fivefold cultures were started with 10 ml, 1 ml, and 0.1 ml in 90 ml, 9 ml, and 9 ml of MK, respectively. The Most Probable Number (MPN) determination was carried out with the aid of the computation tables in "Standard Methods" (10).

TABLE I.—Salmonella species present in Treatment Plant Influent (I) and Effluent (E)

Type	Week					
	1		2		3	
	(I)	(E)	(I)	(E)	(I)	(E)
S. anatum	−	−	−	−	+	−
S. bareilly	+	+	+	−	+	+
S. california	+	−	−	−	−	−
S. dublin	−	−	−	−	−	+
S. give	+	+	+	+	−	−
S. heidelberg	−	−	+	+	−	−
S. manchester	−	−	+	+	+	+
S. meleagridis	−	−	+	−	−	−
S. minnesota	+	−	−	−	−	−
S. montevideo	−	+	−	−	−	−
S. muenchen	−	+	−	−	−	−
S. newington	+	+	−	−	−	−
S. oranienburg	+	−	−	+	+	+
S. panama	+	+	+	+	+	+
S. paratyphi B	+	+	+	+	+	+
S. stanley	−	+	+	+	−	−
S. typhimurium	+	+	+	+	+	+
S. virchow	−	+	−	−	−	−
S. worthington	+	+	+	−	+	+

Note: Samples taken once/wk for 21 days.

Results

Salmonella was regularly found both in the influent and the effluent of this wastewater treatment plant. Three of the 12 salmonella types most common in man in the Netherlands (11) were found in each sample, namely, *S. typhimurium, S. panama,* and *S. paratyphi B* (Table I).

Isolation of salmonella from thoroughly polluted material is exceedingly difficult. This is why several samples of 10 ml must be found false negative if salmonella is isolated from 5, 5, and 2 samples each out of 10 samples with 0.1 ml, 1 ml, and 10 ml influent (Table II). This is probably ascribed, not to the relative concentration of accompanying flora (i.e., the constant ratio contaminants, salmonella), but

TABLE II.—Salmonella Growth in Enrichment Media (37°C Incubation)

Week	Salmonella Positive Samples Present Per 10 Examinations									
	MK MEDIA*						CS MEDIA†			
	0.1 ml in 9 ml MK		1 ml in 9 ml MK		10 ml in 10 ml MK		10 ml in 1 ml CS		100 ml in 10 ml CS	
	Influent	Effluent	Influent	Effluent	Influent	Effluent	Influent	Effluent	Influent	Effluent
1	5	1	5	0	2	7	6	8	9	8
2	2	1	7	1	6	7	3	5	5	5
3	4	2	8	7	3	2	4	7	8	5

* Muller–Kauffman.
† 10-fold concentrate of selenite broth according to Leifson.

to the absolute level of the number of contaminants. The fact that salmonellae are more readily isolated from effluent than from influent supports this hypothesis. However, the results of an examination of 10 samples of 0.1 ml of influent during three samplings (5, 2, and 4 positive samples, respectively) warrant the conclusion that, according to a Poisson distribution, at least 200 salmonella germs should be present per 100 ml influent. In the three samplings of effluent, 1, 1, and 2 samples of the 10 samples of 0.1 ml were positive. This suggests the presence of at least 100 salmonella germs per 100 ml of effluent.

TABLE III.—Most Probable Number (MPN) Salmonella utrecht in Effluent after Addition of 3 × 10¹¹ Organisms to Treatment Plant Influent at 9:00 AM

Sampling Time	Number of Positive Samples/5 examinations			MPN/ 100 ml
	10 ml	1 ml	0.1 ml	
10:00 AM	3	0	0	7.8
10:30 AM	3	0	0	7.8
11:00 AM	5	0	0	23
11:30 AM	5	5	0	240
12:00 NOON	5	2	0	49
12:30 PM	4	3	0	27
1:00 PM	4	4	2	47
1:30 PM	4	4	1	40
2:00 PM	5	3	1	110
2:30 PM	5	3	4	210
3:00 PM	5	3	1	110
3:30 PM	4	1	1	21
4:00 PM	5	1	0	33
4:30 PM	5	3	1	110
5:00 PM	2	1	0	6.8
5:30 PM	4	0	1	17
6:00 PM	5	3	0	79
6:30 PM	4	1	1	21
7:00 PM	3	1	0	11
7:30 PM	5	2	0	49
8:00 PM	4	0	0	13
8:30 PM	4	1	0	17
9:00 PM	4	1	0	17
9:30 PM	5	0	0	23
10:00 PM	3	0	0	7.8
10:30 PM	2	0	0	4.5
11:00 PM	5	0	0	23

The basic assumption in the loading test was that the distribution and passage of *S. utrecht* in the wastewater would be comparable to those of Na-24. This means that within the first 6 hr of introduction of 3×10^{11} *S. utrecht* germs, a maximum of 3×10^{3} germs/100 ml of effluent should be recoverable. In fact, the maximum value found during this period was 240 germs/100 ml; this means a 90 percent reduction in the number of *S. utrecht* germs.

In 10.5 hr (during which 90 percent of the Na-24 disappeared from the plant), an average of 60 *S. utrecht* germs/100 ml were recovered from the effluent. During this time, 10.5 (time) $\times$ 1,500 $\times 10^{4}$ (amount of effluent per hour expressed in 100 ml) $\times$ 60 germs $= 10^{10}$ germs *S. utrecht* escaped with the effluent. This confirms that the eliminating action of the wastewater treatment plant described amounts to one decimal place.

Conclusion

When 3×10^{11} *S. utrecht* germs were added to influent, an average of 60 germs/100 ml effluent were recovered in 10.5 hr. If this is compared with the amounts of salmonella germs normally found in the effluent (100 germs/100 ml), it can be maintained that some 10^{10} salmonella germs continuously enter the plant per hour, while some 10^9 salmonella germs are released from the plant every hour with the effluent (in 1,500 cu m) into publicly accessible water. In view of this situation, the feasibility of disinfection of effluent merits consideration.

Acknowledgment

The system detention time study was carried out by R. van Dongen, Radiation Research Laboratory, National Institute of Public Health, Utrecht.

References

1. Kampelmacher, E. H., Guinée, P. A. M., and Clarenburg, A., "Salmonella Organisms Isolated in the Netherlands During the Period 1951 to 1960." *Zbl. Bakt. I. Abt. Orig.*, **185**, 490 (1962).
2. Guinée, P. A. M., Kampelmacher, E. H., Hofstra, K., and Keulen, A. van, "Salmonellae in Healthy Cows and Calves in the Netherlands." *Zbl. Vet.-Med., B.*, **11**, 728 (1964).
3. Schothorst, M. van, Guinée, P. A. M., Kampelmacher, E. H., and Keulen, A. van, "Prevalence of Salmonellae in Poultry in the Netherlands." *Zbl. Vet.-Med., B.*, **12**, 422 (1965).
4. Guinée, P. A. M., Kampelmacher, E. H., Valkenburg, J. J., "Salmonella Isolations in the Netherlands, 1961–1965." *Zbl. Bakt. I. Abt. Orig.*, **204**, 476 (1967).
5. Schaaf, A. v. d., and Atteveld, J. C., "Voorkomen van Salmonellae in het Effluent van Moderne Inrichtingen van Rioolwaterzuivering." *Geneesk Gids.*, **43**, 57 (1965).
6. Kampelmacher, E. H., "Comparative Studies in the Isolation of Salmonella from Minced Meat in Five Laboratories." *Zbl. Bakt. I. Abt. Orig.*, **204**, 100 (1967).
7. Taylor, W. L., "Isolation of Salmonellae from Food Samples." *Appl. Microbiol.*, **9**, 487 (1961).
8. Harvey, R. W. S., and Thomson, S., "Optimum Temperature of Incubation for Isolation of Salmonellae." *Monthly Bull. Minist. Health Lab. Serv.*, **12**, 149 (1953).
9. Edel, W., and Kampelmacher, E. H., "Comparative Studies on Salmonella Isolation in Eight European Laboratories." *Bull. World Health Org.*, **39**, 487 (1968).
10. "Standard Methods for the Examination of Water and Wastewater." 10th Ed., Amer. Pub. Health Assn., New York, N. Y., 382 (1955).
11. Kampelmacher, E. H., "Laboratorium voor Zoönosen." *Verslagen en Mededelingen betreffende de Volksgezondheid*, **20**, 32 (1968).

Additional References

Kampelmacher, E. H., Guinée, P. A. M., Hofstra, K., and Keulen, A. van, "Further Studies on Salmonella in Slaughterhouses and in Normal Slaughter Pigs." *Zbl. Vet.-Med., B.*, **10**, 1 (1963).

Schligmann, R., and Rettler, R., "Enteropathogens in Water with Low *Esch. Coli* Titer." *Jour. Amer. Water Works Assn.*, **57**, 1572 (1965).

Tompkin, R. B., Weiser, H. H., and Malaney, G. W., "Salmonella and Shigella Organisms in Untreated Farm Pond Water." *Jour. Amer. Water Works Assn.*, **55**, 592 (1963).

Weibel, S. R., Dixon, F. R., Weidner, R. B., and McCabe, L. J., "Waterborne-Disease Outbreaks, 1946–60." *Jour. Amer. Water Works Assn.*, **56**, 947 (1964).

REDUCTION OF *SALMONELLA* IN COMPOST IN A HOG-FATTENING FARM OXIDATION VAT

E. H. Kampelmacher and Lucretia M. van Noorle Jansen

In recent years the number of large hog-fattening farms, where frequently thousands of animals are fattened, has increased greatly. On these farms, the disposal of urine and feces, which in such large quantities can no longer be used for fertilization, constitutes a serious problem. In order to deal with it by means of purification, oxidation vats have been constructed and oxidation ditches dug on these farms.

A description is given in this paper of experiments carried out with a small experimental oxidation vat suitable for 160 hogs and installed on a hog-fattening farm in Wageningen by the Instituut Landbouw en Bedrijfsgebouwen (Agriculture and Farm Buildings Institute). The study was performed to determine whether bacteriological reduction in general and elimination of *Salmonella* in particular occur, and if so, to what extent.

Material and Methods

Description of the Plant

The 30-cu m oxidation vat described here was installed on a farm where 150 fattening hogs are housed in 16 cages in a stable. The feces and urine are collected through grids in a manure cellar. From this cellar, approximately 0.6 cu m of manure mixed with approximately 1.9 cu m of water is pumped into the oxidation vat, which is adjacent to the stable (Figures 1 and 2). A rotor in the vat provides aeration. The diameter of the aeration plant is 75 cm (Figure 3), and the rotor has four speeds. Only the first two speeds are used in this vat because at higher speeds the mixture is agitated too vigorously. Each morning the plant is switched off for 1 hr to allow the mixture to settle. Then an aperture near the upper margin of the vat is opened so that the effluent can flow into a ditch near the farm (Figure 4). After the aperture has been closed again, manure is once more pumped up from the manure cellar and is then mixed with water and aerated for the rest of the day and the succeeding night.

Bacteriological Examination

This part of the investigation consisted of qualitative determination of *Salmonella* in the feces, manure, active mixture, and effluent and ditch water, and quantitative determination of the aerobic bacteria, enterobacteriaceae, and particularly *Salmonella* in the active mixture and the effluent and ditch water. For the feces study, two random fresh samples of approximately 100 g were collected from each cage, that is, a total of 32 samples of feces. Of the other products, 10 samples of 100 ml each were collected of compost on its way up from the manure cellar during pumping, of the active mixture immediately after switching off the aeration plant, of the effluent during outflow, and of the ditch water before the effluent drained into it.

Muller-Kauffmann (MK) tetrathionate broth was used as the enrichment

FIGURE 1.—Pipe through which manure is pumped from the cellar into the oxidation vat.

FIGURE 3.—Oxidation vat with aeration rotor.

medium in the isolation of the salmonellae. After incubation at 43°C for 18 to 24 and 45 to 48 hr, the material was spread on 14-cm diam brilliant green (BG) phenol red agar plates. The plates were then incubated at 37°C for 18 to 20 hr. Suspected colonies were reinoculated onto triple sugar iron agar (DIFCO) and in lysine decarboxylase medium. If a positive biochemical result was obtained, serological typing was carried out at the National Salmonella Center. For the qualitative examination, 10 g or 10 ml of the 100-g or 100-ml samples were incubated with 150 ml MK, while for the quantitative examination, 1 ml and 0.1 ml were incubated with 10 ml MK. The most probable number (MPN) calculation was carried out with the aid of previously published tabulated data.[1] The number of aerobic bacteria and the number of enterobacteriaceae were determined by duplicate spreading of 0.1 ml of 10^{-1}, 10^{-3}, and 10^{-5} dilutions on blood plates and on violet red bile glucose agar plates that were then incubated for 24 hr at 37°C. The mean value from 24 determinations of 12 samples was calculated and recorded.

FIGURE 2.—Influent pipe and oxidation vat.

FIGURE 4.—Effluent pipe from oxidation vat to ditch.

TABLE I.—Bacterial Counts in 12 Samples of Active Mixture, and of Effluent and Ditch Water*

Sample	Average Count and Deviation (no./ml)	
	Aerobic Bacteria	Enterobacteriaceae
Active mixture	$6.4 \cdot 10^6$ ($1.4 \cdot 10^6$–$2.8 \cdot 10^7$)	$3.8 \cdot 10^4$ ($1.1 \cdot 10^4$–$1.3 \cdot 10^5$)
Effluent	$2.2 \cdot 10^5$ ($1.8 \cdot 10^4$–$2.6 \cdot 10^6$)	$8.3 \cdot 10^2$ ($1.0 \cdot 10^2$–$7.0 \cdot 10^3$)
Ditch water	$1.5 \cdot 10^6$ ($3.0 \cdot 10^5$–$7.9 \cdot 10^6$)	$2.6 \cdot 10^4$ ($5.8 \cdot 10^3$–$1.2 \cdot 10^5$)

* 24 counts/sample.

Results

In the course of the investigation, 12 bacterial counts were made of samples of the active mixture and of the effluent and ditch water. The average values obtained are listed in Table I.

Table II lists the results of all studies for the presence of *Salmonella* in samples of feces, manure, active mixture, and effluent and ditch water.

In Table III results are summarized for five *Salmonella* MPN determinations of material collected during three different days.

Discussion

The bacterial counts revealed that the processing in the oxidation vat reduces the aerobic flora reasonably well. From the more detailed MPN determinations of the *Salmonella* counts, it could be concluded that the *Salmonella* reduction amounted to two decimal points, which is in accordance with the requirements of *E. coli* in treatment plants for wastewater.

In practice, this means that when the *Salmonella* infection of the animals is not too severe, the number of *Salmonella* organisms that escape in the wastewater will be so small that the possibility of human or animal infection by means of surface water contamination can be regarded as remote (Table II). On the other hand, once the active mixture achieves a content of 10^2 to 10^3 *Salmonella* bacteria (which occurred after September 9), the possibility exists, even after a 100-fold reduction achieved by the plant, of *Salmonella* bacteria being constantly present in the effluent in numbers that can be viewed as undesirable (Table III).

In summary, it may be stated that for a treatment plant of the type described here, the number of *Salmonella* bacteria in the effluent depends entirely on the original *Salmonella* count—in this case on the severity of the infections in the hogs. As soon as some sanitation is achieved on hog-fattening farms in the Netherlands, the number of hogs infected with *Salmonella* will fall considerably.[2] To this end, treatment of feces and urine by means of oxidation vats and oxidation ditches on large fattening farms will probably prove most satisfactory. In the meantime, however, it may be considered appropriate to decontaminate the effluent from such plants by other means, (such as chlorination) especially because the effluent is usually sluiced out to surface water. In this way the infection of animals in pastures and of humans in recreation areas may be prevented. Such infection has previously been reported in the literature.[3]

Summary

The reduction of *Salmonella* in compost in an experimental oxidation vat on a hog-fattening farm was shown to be approximately 100-fold. As long as the excretion of *Salmonella* in the feces of the hogs remains low, only small numbers of *Salmonella* bacteria will be sluiced out with the effluent; thus,

TABLE II.—*Salmonella* Isolations from Feces, Manure, Active Mixture, Effluent and Ditch Water

Test Number	Date	Feces* Number of Positive Samples	Feces* Serotype‡	Manure† Number of Positive Samples	Manure† Serotype‡	Active Mixture† Number of Positive Samples	Active Mixture† Serotype‡	Effluent† Number of Positive Samples	Effluent† Serotype‡	Ditch Water† Number of Positive Samples	Ditch Water† Serotype‡
1	5/29		n.p.		n.p.		—		n.p.		—
2	6/9		n.p.		—		—		—		—
3	6/23	1	*S. cubana*		n.p.		n.p.		n.p.		n.p.
4	7/7		—		n.p.		n.p.		n.p.		n.p.
5	7/15	18	*S. anatum*		n.p.		n.p.		n.p.		n.p.
6	7/30	5	*S. typhimurium*	8 {4	*S. anatum*	3	*S. typhimurium*	3	*S. typhimurium*		—
				4	*S. typhimurium*						
7	8/12	3 {2	*S. anatum*	3 {2	*S. tennessee*	1	*S. typhimurium*		—		—
		1	*S. gtve*	1	*S. typhimurium*						
8	8/20	2 {1	*S. enteritidis*	1	*S. eimsbuettel*		—		—		—
		1	*S. typhimurium*								
9	8/27	3 {2	*S. anatum*	1	*S. eimsbuettel*		—		—		—
		1	*S. typhimurium*								
10	9/3	11 {4	*S. anatum*	10	*S. panama*	10	*S. panama*		—		—
		6	*S. panama*								
		1	*S. typhimurium*								
11	9/9	30 {5	*S. agona*	10	*S. panama*	10	*S. panama*	10	*S. panama*		—
		23	*S. panama*								
		2	*S. senftenberg*								
12	9/12	32 {2	*S. agona*	10	*S. panama*	10	*S. panama*		—	8	*S. panama*
		30	*S. panama*								
13	9/16	31 {30	*S. panama*	10	*S. panama*	10	*S. panama*	3	*S. panama*	2	*S. panama*
		1	*S. typhimurium*								
14	9/18	32	*S. panama*	10	*S. panama*	10	*S. panama*	6	*S. panama*	3	*S. panama*
15	9/23	29 {3	*S. anatum*	10	*S. panama*	10	*S. panama*	10	*S. panama*	10	*S. panama*
		26	*S. panama*								
16	9/30	21 {1	*S. derby*	9 {7	*S. derby*	10 {1	*S. derby*	7	*S. panama*	10	*S. panama*
		4	*S. eimsbuettel*	2	*S. panama*	9	*S. panama*				
		16	*S. panama*								
17	10/7	20 {3	*S. agona*	10	*S. panama*	10	*S. panama*	1	*S. panama*	9 {1	*S. agona*
		2	*S. infantis*							8	*S. panama*
		15	*S. panama*								
18	10/14	17 {3	*S. anatum*	10	*S. panama*	10	*S. panama*	1	*S. panama*	10 {1	*S. anatum*
		2	*S. derby*							9	*S. panama*
		2	*S. heidelberg*								
		4	*S. infantis*								
		5	*S. panama*								
		1	*S. senftenberg*								

* Per test, 32 samples of 10 g each. † Per test, 10 samples of 10 ml each. ‡ n.p. = not performed; — = negative.

TABLE III.—*Salmonella* MPN Determinations

Date	MPN (no./100 ml)		
	Active Mixture	Effluent	Ditch Water
9/12	240	<1.05	16
9/18	730	9.16	3.57
9/30	1,600	12	25
10/7	63	1.05	23
10/14	33	1.05	75

the possibility of human and/or animal infection in surface water may be regarded as very slight. As soon as either the number of hogs or the number of excreted bacteria, or both, increases, the plant can no longer adequately reduce the number of salmonellae to the extent that the effluent can be sluiced out into open water without the danger of human or animal infection. Therefore, it is recommended that adequate chlorination of the effluent of such plants be considered.

References

1. Committee on Bathing Beach Contamination, P.H.L.S. "Sewage Contamination of Coastal Bathing Waters in England and Wales." *Jour. Hyg. Camb.*, **57**, 435 (1959).
2. Edel, W., *et al.*, "Effect of Feeding Pellets on the Prevention and Sanitation of *Salmonella* Infections in Fattening Pigs." *Zbl. Vet. Med. B.* **17**, 730 (1970).
3. Strauch, D., and Parráková, E., *Mitt. deutsch. Landwirt.-Ges.*, **40**, 1256 (1969).

Reduction of bacteria in sludge treatment

E. H. KAMPELMACHER AND LUCRETIA M. VAN NOORLE JANSEN

BECAUSE THE VOLUME of wastewater reaching treatment plants is constantly rising, the quantity of sludge is also increasing. In conventional plants this sludge, which has a very high water content (95 to 99 percent), is allowed to ferment and is then led to sludge beds or fields to dry. The drying depends on climatic conditions such as the temperature and rainfall. However, even under very favorable meteorological conditions, such as a hot summer or a long period without much rain, this means that the proportion of dry substances in the sludge increases from an average of 3 percent before drying to some 30 percent, at best, after months of drying. Only the summer months are suitable for this process, which is time-consuming and takes up a great deal of space. Consequently, other sludge-drying methods have been developed, for example, filtration of sludge over vacuum filters, followed by the addition of chemicals and passage of the mud over filter drums, where part of the water is extracted. The end product obtained in this manner consists of 25 to 30 percent dry substance. This material can be transported without difficulty and may be used, for instance, in landscaping for fill material. It can also be incinerated—a process already practiced in some places. Earlier personal investigations and reports from the literature from the Netherlands and other countries [1,2] indicate that the sludge and the product in which it is often incorporated, namely, compost, are often severely contaminated with *Salmonella* microorganisms. A report is presented in this paper of an investigation in three sludge treatment plants in the Netherlands, which was made to determine whether, in addition to the above-mentioned advantages, the sludge treatment process also leads to a reduction in the number of aerobic microorganisms, particularly *Salmonella*.

DESCRIPTION OF THE PLANTS

Apeldoorn treatment plant. The sludge from this treatment plant, which consists of approximately 75 percent primary components and 25 percent surplus active sludge, is collected in a storage basin and then led into a drying tank, with the addition of approximately 2 percent ferric chloride and 10 percent lime, in order to improve flocculation. Thereafter, the sludge is led over a drying drum covered with a filtering blanket of polypropylene POPR 929, through which the water is sucked out by means of a vacuum created in the drum. The so-called sludge cake produced in this way is transported on a Jacob's ladder to containers, which in the Apeldoorn plant are emptied into a pit between alternating layers of sand. The drained water is returned to the wastewater purification plant—as is the water (tap water) used in regular washing of the filtering blanket. The primary sludge contains approximately 8 percent dry substance and the end product approximately 27 to 30 percent.

Mierlo treatment plant. The primary sludge is pumped from the wastewater treatment plant in Eindhoven to Mierlo where it is collected in a storage basin. The amounts of lime and ferrous sulfate added at this point to condition the sludge depend on its composition: usually the ratio ferrous sulfate:lime is 1:6, but when the sludge possesses good filtering qualities, less ferrous sulfate is added. The amounts added are calculated in retrospect/1,000 kg of sludge cake, and they vary between 30 and 40 kg lime and 5 and 25 kg ferrous sulfate. After the addition of the chemicals, the sludge is led over a

TABLE I.—Mean Values and Variations in the Bacterial Counts in Wet and Dry Sludge from the Apeldoorn Plant

Type of Plate*	Mean and Range of Count	
	Wet Sludge†	Dry Sludge†
Blood	3×10^6	2×10^4
	3×10^5–3×10^7	7×10^3–7×10^4
Negram	2.10^5	7.10^2
	3×10^4–1×10^6	5×10^1–9×10^3
Endo	2.10^6	1.10^3
	2×10^5–3×10^7	1×10^2–8×10^3

* Plates were incubated for 24 hr at 37°C under aerobic conditions.

† 26 samples of 1 g.

filter comprising a vacuum drum with a surface of fine wires. Before the drying, the sludge contains approximately 4 percent of dry substance and after drying approximately 25 percent. The plant at Mierlo contains two drums, which on an average produce 75 tons of dried sludge/day. Production can be increased to a maximum of 100 tons/day.

Maastricht treatment plant. The sludge that settles in the after-precipitation of the wastewater treatment plant may, if necessary, be returned to the aeration tanks so that a given concentration of sludge can be maintained. After passage through pre-precipitation tanks, aeration tanks, and after-precipitation tanks, the sludge is collected in the sludge-thickening plant where it is thickened and moved centrifugally by rotating blades to the center of the tanks, and the excess water is returned to the pre-precipitation tanks. Subsequently, the sludge is led to the chemical mixer. The daily additions of lime and ferrous sulfate are determined by laboratory titrations. Particularly when secondary sludge is treated, the amounts of chemicals must be determined accurately because they vary greatly. These amounts are calculated per unit volume of dried sludge and are roughly in agreement with the figures given for the Mierlo plant. Here, again, the sludge is led over a wire-surfaced vacuum filter. The proportion of dry substance of the "thickened" sludge is approximately 9.3 percent, while the dried sludge contains approximately 26 percent. About 300 to 400 cu m of dried sludge is produced each month.

Materials and Methods

Bacteriological examination. The bacterial counts were determined in 1 g of wet sludge or in dry sludge, as the case may be. These quantities were dissolved in 10 ml physiological saline, and a number of dilutions were then prepared. Each dilution (0.1 ml) was spread on a blood plate to determine the number of aerobic microorganisms, and a nalidixic acid-agar plate (Negram plate) was used to identify Gram-positive bacilli and cocci and an endoagar plate for the number of enterobacteriaceae. The count was made after the plates had been incubated for 24 hr at 37°C under aerobic conditions.

***Salmonella* study.** Of the samples of wet sludge that had been collected at random, 10-g quantities were mixed with 100 ml of Muller-Kaufmann's tetrathionate broth. Of the samples of dry sludge, also collected at random, 2.5-g quantities were mixed with 100 ml of the broth. Subsequently, the jars were placed in a water bath at 45°C for 15 min in order to heat

TABLE II.—*Salmonella* Isolated from Wet and Dry Sludge from the Apeldoorn Plant

Variable	Value	
	Wet Sludge	Dry Sludge
Volume of sample (g)	10	2.5
Total number of samples	72	144
Salmonella negative	25	137
Salmonella positive	47 (65.3%)	7 (4.9%)
Type distribution:		
S. anatum	1	—
S. derby	2	—
S. eimsbuettel	2	—
S. enteritidis	3	—
S. litchfield	9	2
S. newport	4	1
S. reading	3	—
S. san diego	4	—
S. stanley	3	—
S. tennessee	2	—
S. typhimurium	14	4

the fluid as quickly as possible to the desired temperature of 43°C. This was followed by incubation in an incubator at 43°C. After 20 hr and again after 48 hr, some of the enrichment media was spread on brilliant green phenol red agar plates (BGA) with a diameter of 14 cm. Suspected colonies were subjected to further biochemical and serological examination.[3]

Results

Apeldoorn plant. Samples from this plant were taken on 12 occasions during the months of June and July. The results of the bacterial counts are summarized in Table I and show that the processing to dry sludge leads to a reduction in the number of aerobic bacteria of the order of two decimal points. This is also true of the Gram-positive flora, while the reduction in the Gram-negative flora amounted to three decimal points. The results of the *Salmonella* study are given in Table II. While 65.3 percent of the samples of the wet sludge proved contaminated, only 4.9 percent of the samples of the dry sludge contained *Salmonella* microorganisms.

Mierlo plant. Samples from this plant were taken on 12 occasions during the months of June and July. The results of the bacterial counts are given in Table III. Here again, a reduction of the order of two decimal points was found in the aerobic bacteria and in the Gram-positive flora. The reduction in the Gram-negative flora was again of the order of two decimals. The reduction in the number of *Salmonella* positive samples was of the same order as in the Apeldoorn plant, namely, 59 percent positive samples of wet sludge as against 5 percent positive samples of dry sludge. The results of the *Salmonella* study are shown in Table IV.

Maastricht plant. Samples from this wastewater plant were taken on 12 occasions during the months of June and July. The results of the bacterial counts are listed in Table V. The most interesting findings were the marked reduction (of the order of three decimals) in the total aerobic flora and an even greater reduction (of the order of four decimals) of the Gram-negative flora. All the samples of wet sludge proved to be contaminated with *Salmonella,* while 13.5 percent of the samples of dry sludge were positive.

The results of the *Salmonella* study are listed in Table VI.

TABLE III.—Mean Values and Variations of Bacterial Counts in Wet and Dry Sludge from the Mierlo Plant

Type of Plate*	Mean and Range of Count		
	Wet Sludge†	Dry Sludge	
		Filter I‡	Filter II†
Blood	4×10^6	2×10^4	3×10^4
	6×10^5–3×10^7	3×10^3–1×10^5	4×10^3–3×10^5
Negram	8.10^4	7.10^2	3.10^2
	7×10^3–1×10^6	5×10^1–8×10^3	6×10^1–2×10^3
Endo	8.10^5	2.10^3	2.10^3
	5×10^4–1×10^7	2×10^2–1×10^4	2×10^2–2×10^4

* Plates were incubated for 24 hr at 37°C under aerobic conditions.
† 26 samples of 1 g.
‡ 16 samples of 1 g.

TABLE IV.—Salmonella Isolated from Wet and Dry Sludge from the Mierlo Plant

Variable	Value		
	Wet Sludge	Dry Sludge	
		Filter I	Filter II
Volume of sample (g)	10	2.5	2.5
Total number of samples	66	41	72
Salmonella negative	27	39	72
Salmonella positive	39 (59%)	2 (5%)	—
Type distribution:			
S. anatum	5	—	—
S. brandenburg	2	—	—
S. bredeney	1	1	—
S. derby	4	—	—
S. eimsbuettel	3	—	—
S. heidelberg	2	—	—
S. infantis	2	1	—
S. litchfield	1	—	—
S. muenchen	1	—	—
S. newport	3	—	—
S. poona	2	—	—
S. pretoria	1	—	—
S. tennessee	3	—	—
S. thompson	1	—	—
S. typhimurium	6	—	—
S. worthington	1	—	—
Salmonella bioch. positive	1	—	—

TABLE V.—Mean Values and Variations of Bacterial Counts in Wet and Dry Sludge from the Maastricht Plant

Type of Plate*	Mean and Range of Counts	
	Wet Sludge†	Dry Sludge†
Blood	3×10^{8}	1×10^{5}
	8×10^{7}–8×10^{8}	1×10^{4}–8×10^{5}
Negram	3.10^{6}	5.10^{3}
	8×10^{5}–8×10^{6}	2×10^{2}–1×10^{5}
Endo	2.10^{8}	2.10^{4}
	6×10^{7}–4×10^{8}	3×10^{3}–1×10^{5}

* Plates were incubated for 24 hr at 37°C under aerobic conditions.
† 16 samples of 1 g.

Discussion

The process of drying sludge in treatment plants, as described in the introduction, results in a marked reduction in bacterial content of the end product. The number of aerobic bacteria is reduced by the order of two to three decimals. No significant difference could be observed between the rate of destruction of the various bacterial species mentioned above: the results in the three plants were nearly the same with respect to reduction in the aerobic flora. A distinct reduction in *Salmonella* could be observed, which was manifested in the proportions of the positive samples in the wet and dry sludge. While the percentages of positive samples of the end product were nearly the same in Apeldoorn and Mierlo, more positive samples of dry sludge were found in Maastricht. In this connection, it should be noted that 100 percent of the samples of wet sludge in Maastricht were contaminated with *Salmonella.*

It may be assumed that the reduction in the numbers of bacteria was the result of the addition of chemicals, producing a strongly alkaline medium and possessing a bactericidal action.[4] The bactericidal effect of the drying as such may be ignored because the a_w of dried sludge still amounted to approximately 0.98, as was shown by a number of estimations. In summary, it may be stated that the process of sludge treatment with vacuum drums presents not only the advantages mentioned in the introduction, but also causes a considerable reduction in aerobic bacteria, particularly in enterobacteriaceae including *Salmonella.* It is especially the latter effect which, with a view to further processing of dry sludge, is of public health importance.

Summary

The filtration of sludge, carried out with the aid of ferric chloride or ferrous sulfate and lime and with the aid of vacuum filters, yields an end product in which the number of bacteria is considerably reduced. In the sludge treatment plants at Apeldoorn and Mierlo, a reduction of the order of two decimals was found for the aerobic bacteria and of the order of two to three decimals for the content of enterobacteriaceae. In the plant in Maastricht, the reduction in the number of aerobic bacteria was of the order of three decimals and that of enterobacteriaceae of four decimals. In all three plants, *Salmonella* contamination of the dry sludge was significantly less than that of the wet sludge.

TABLE VI.—Salmonella Isolated from Wet and Dry Sludge from the Maastricht Plant

Variable	Value	
	Wet Sludge	Dry Sludge
Volume of sample (g)	10	2.5
Total number of samples	48	96
Salmonella negative	—	83
Salmonella positive	48 (100%)	13 (13.5%)
Type distribution:		
S. eimsbuettel	1	—
S. infantis	12	—
S. meleagridis	4	—
S. muenchen	1	—
S. panama	13	4
S. saint paul	1	—
S. thompson	1	—
S. typhimurium	9	8
S. worthington	6	—

ACKNOWLEDGMENTS

Credits. The authors thank D. A. A. Mossel, Central Institute for Nutrition and Food Research, Zeist, who carried out estimations of a_w.

REFERENCES

1. Knoll, K. H., "Hygienic Aspects in the Treatment and Removal of Sludge and Other Solid Wastes." *Schweiz. Zeits Hydrol.* (Switz.) **XXVI,** 693 (1964).
2. van der Schaaf, A. "Incidence of Salmonellae in Effluent of Modern Sewage Treatment Plants." *Geneesk. Gids.,* **43,** 3, 57 (1965).
3. Guinée, P. A. M., *et al.,* "Salmonellae in Healthy Cows and Calves in the Netherlands." *Zbl. Vet.-Med.,* **11,** 728 (1964).
4. Doyle, C. B., "Effectiveness of High pH for Destruction of Pathogens in Raw Sludge Filter Cake." *Jour. Water Poll. Control Fed.,* **39,** 1403 (1967).

Pollution of Water with Organic and Inorganic Chemicals

OIL–A NEW YORK STATE POLLUTION PROBLEM

Willard A. Bruce and Irving Grossman

One of the most publicized and controversial types of pollution is that caused by oil and other petroleum products. Whereas wastes from municipalities and industries can be successfully collected, treated, and rendered harmless before discharge to streams, the multitude of small spills or releases of petroleum products occurring yearly cause property damage, impair water quality, and destroy wildlife. In New York state approximately 200 incidents involving the release of petroleum products to waterways are reported yearly. Although dwarfed in magnitude by the Santa Barbara runaway well incident or the Torrey Canyon disaster,[1] the total damage from these intermittent incidents is great and the interference with normal water usage a matter of public concern. This paper discusses the efforts initiated in New York to cope with oil pollution and briefly mentions the national and international aspects of the problem.

Oil and Petroleum Products

Over 6.5 bil gal (24.6 mil cu m) of gasoline were consumed in New York during 1969. Constituting one of the largest sources of tax revenue, gasoline is transported into the state by tank trucks, ships, barges, and underground pipelines.

Lubricating oils used in the state add up to several thousand tons per year. Crude oil in relatively small quantities is obtained from wells in Allegany, Cattaraugus, and Steuben counties. There are two refineries in the state; one processes crudes in the Buffalo vicinity, and the other refines used oils in the Syracuse area. Almost all of the gas and petroleum products used in New York are obtained by processing crudes in out-of-state establishments and shipping the products into the state.

Table I is a tabulation showing the fuel oils commonly transported and consumed in the state.

Kerosene is refined petroleum distillate suitable for use as an illuminant. Synonymous terms are lamp oil, burning oil, illuminating oil, and range oil when the product is used in space heaters. When sold as a range oil for space heaters, it is frequently identified as No. 1 distillate fuel oil.

Fuel oils are liquid or liquefiable petroleum products burned for the generation of heat in a furnace or firebox or for the generation of power in an engine. These do not include oils with a flash point below 100°F (38°C) and oils burned in cotton- or wool-wick burners.

Residual fuel oils are topped crude petroleum or viscous residuums obtained in refinery operations. Commercial grades of burner fuel oils Nos. 5 and 6 are residual oils and include Bunker fuels. The latter, used by ships and industry for large-scale heating installations, are similar in requirements to No. 6 grade fuel oil. These fuel oils and gasoline are most commonly involved in oil pollution incidents occurring during periods of transfer and transport.

Jet fuel, which is a refined form of kerosene similar in characteristics and physical properties to No. 1 fuel oil, was excluded from the tabulation.

Figures are not available for the total quantities of petroleum products handled or consumed in this state. However, an indication of quantities handled on the Barge Canal System during the 1969 shipping season is shown in Table II.

The quantities indicated in Table II do not include petroleum products consumed in the New York metropolitan area or Hudson River valley area as far north as Troy. An interesting fact is that 67 percent of all cargo transported on the Canal System is petroleum and petroleum products.

TABLE I.—Standard Fuel Oils*

Grade	Comments
No. 1	A distillate oil intended for vaporizing pot-type burners and other burners requiring this grade of fuel
No. 2	A distillate oil for general purpose domestic heating for use in burners not requiring No. 1 fuel oil
No. 4	Preheating not usually required for handling or burning
No. 5 (Light)	Preheating may be required depending on climate and equipment
No. 5 (Heavy)	Preheating may be required for burning and in cold climates heat may be required for handling
No. 6	Preheating required for burning and handling

* American Petroleum Institute, New York, N. Y.

Spills and Releases

Most oil pollution problems are caused by accidents or incidents that involve petroleum products in transport via pipeline, tank truck, ships, or barges. Waterway incidents have resulted from ruptured holds and from oily ballast water discharges while the carrier was under way.

Shoreline incidents during periods of ship-to-shore or shore-to-ship transfers have been traced to improper coupling and uncoupling procedures, ruptured hose lines, negligent operation of valves and shut-offs, and overpumping.

The vast quantity of petroleum marketed and transported each year is stored before consignment to the consumer at many tank farms along major waterways. Leaking tanks and faulty valves have resulted in frequent releases of large quantities of petroleum products to groundwater and surface water. At truck platforms there have been many cases of spills caused by poor housekeeping, overloading of

TABLE II.—Tonnage of Petroleum and Its Products Handled in 1969 on the New York State Barge Canal System*

Product	Erie Division	Champlain Division	Oswego Division	Total System
Gasoline and motor fuel	108,317	447,418	9,479	565,214
Kerosene	14,937	78,798	—	93,735
Jet fuel	24,890	127,061	9,480	161,431
Fuel oil nos. 2 and 3	115,611	496,895	28,719	641,225
Fuel oil nos. 4 and 6	470,174	215,545	2,400	688,119
Lubricating oil and grease	41,411	—	3,300	44,711
Other petroleum products	2,775	—	15,893	18,668
Total petroleum products	778,115	1,365,717	69,271	2,213,103

* New York State 1969 Canal Tonnage Report.
Note: Tons × 0.907 = metric tons.

truck tanks, and rupturing of outlet pipes. Improper platform drainage has contributed oil to waterways, and yard drainage from storage facilities, particularly in New York City, has caused oil pollution problems.

Many incidents in the western part of the state have been attributed to the rupture of transmission lines at stream crossings. Gasoline and fuel oil are conveyed to New York by interstate transmission lines running for miles under the earth's surface to the market areas. Their operation is controlled by a complex automatic system. Malfunctioning of the equipment has resulted in overpumping to storage tanks and subsequent release of gasoline to surface waters.

The western New York-Pennsylvania border area of the state has numerous well-drilling operations where the crude oil is collected in a common pipe system that serves a multiple number of wells. Poor housekeeping and maintenance have resulted in releases of large quantities of crudes to surface waters. Such crudes have the same damaging effects as fuel oils in addition to a high salt content.

Other important sources of petroleum pollution involve the consumer. During delivery, gravity feed from the tank truck to the underground storage tank is often not properly supervised. Overflows from the receiving tanks have discharged to surface waters. Ruptures of storage tanks and pipeline breaks have caused oil pollution spills. Faulty heat exchangers at large institutional, commercial, and office buildings have often leaked for days before the source of oil leakage was discovered and the problem corrected. Leaky underground storage tanks have contaminated groundwater supplies. Poor housekeeping at industrial plants and railroad yards has saturated the ground to the point where oil seeps steadily to waterways and will continue to be a problem for many years.

Used crankcase oil disposal from filling stations is becoming a serious problem because of the changes in the tax structure which makes re-refining of waste oils questionable from an economic standpoint.

Effects of Spilled Products on Environment

Any spill or relase of oil to water causes a problem. Oil is an irritating pollutant; only a small amount makes the most conservative conservationist "bristle." Being more volatile than water and having hazardous ignition and combustion characteristics, there is a potential fire hazard associated with almost any spill, particularly when the spill occurs under a dock, boathouse, or similar structure. Also, the oil discolors water and shoreline areas, destroying the natural beauty of vegetated areas and property. Beaches, marinas, docks, and pleasure craft have suffered extensive damage because of the time-consuming and difficult task of removing the tacky substance.

Although most hydrocarbons have toxicity characteristics capable of killing fish, the major problems are long-term accumulation of oil causing tainting of the meat and the destruction of microscopic plant life. The heavier oils, No. 5, No. 6, and the Bunker fuel oils, will often sink to the bottom when part of the solvent components evaporate. In shallow shoreline areas where fish propagate, this can destroy feed beds and spawning areas. The tacky oil adheres to the bottom and is decomposed slowly by bacteria.

Waterfowl have suffered extensive damage because of contact with oil. The calming effect of oil on water induces ducks to alight on such waters before they will land on adjacent rough waters. Once they alight and come into contact with the oil, the water-shedding property of the feathers is destroyed. This situation renders the birds more susceptible to the elements and enemy predators. In attempting to cleanse itself of the oil, the bird may ingest some of the oil and die.

Spills have affected industrial water supplies. A paper mill was temporarily shut down because the water could not be used. A pharmaceutical manufacturing establishment had to cease operating until its sand filters and intake structure were thoroughly cleaned. Fortunately, because oil is lighter than water, intakes located in lower depths are usually not affected. Occasionally some oil may emulsify with the receiving waters to such an extent that it enters a water intake. Taste and odor could render a supply worthless until the condition is corrected.

Cleanup Techniques

Small Inland Waters

Most oil pollution problems in New York consist of many small isolated spills. The immediate objective is to contain the spill and prevent its spread or transport out of the area by wind or water current. Containment is usually accomplished by using plastic retaining booms or constructing a "hay bale" dam. Once contained, the responsible party must remove and properly dispose of the oil and other associated wastes. Information on available equipment in the area and technical guidance by Health Department personnel is provided. Chemicals to disperse or emulsify the oil are not allowed.

Lakes and Coastal Waters

For spills in large bodies of water such as Lake Ontario or the New York metropolitan coastal waters, the alternatives to be considered include: (*a*) capture or containment, (*b*) sinking, (*c*) burning, (*d*) chemical treatment to disperse or emulsify, or (*e*) letting them drift ashore or out to sea.

Except on very smooth water and under ideal conditions it is nearly impossible to capture or contain oil on wide expanses of water. To obtain, assemble, and apply material such as coal dust, rock dust, sand, or other inert material to sink the oil on wide open expanses of water is extremely difficult. The practicality of burning is uncertain, and there is a need for further testing and development of this technique. Oil in one slick on calm waters, immediately after the spill, may be subject to burning but such conditions are not likely to exist by the time action is taken. Although attempts to burn oil from the Torrey Canyon were unsuccessful, recent reports indicate spilled oil was burned with a fair degree of success in the coastal waters of Sweden.

Chemical treatment of spills far from shore, when it seems that the slick will move to shore, is a method deserving consideration. However, the chemicals must be harmless to natural plant or animal growth in the waters and nontoxic to man and fish life; they should have a reasonable decay period.

If a large spill occurs and meteorological conditions are conducive to moving it out to sea, the best course would be to take no action. Evaporation of the solvent phase plus bacterial action either destroys the oil or causes it to sink.

Cleanup of Shorelines, Equipment, and Beaches

Once oil is spilled near shore areas or blown to shore from the sea, the real damaging effects will occur. The heavier black fuel oils and crude oils leave a sticky film on contact. Some oil may be contained in shallow calm bay areas and arrested until removed. Other oils may be seeded with hay in shallow areas to facilitate removal. Natural driftwood shoals may form a barrier or trap and contain much of the oil, thus facilitating removal. However, in a large spill where the oil floats to shore, the large cleanup operations and the damage may be substantial. Often beach sand must be removed and replaced with clean sand.

Docks and boats must be manually cleaned, although some chemical treatment may be allowable. Rocks and weed beds may have to be left to nature's slow cleaning action.

Prevention and Control

Prevention

In comparison with most pollution problems caused by continuous disposal of wastes, oil pollution usually results from accidents or negligence. Therefore, control of oil pollution must include an educational and cooperative approach. The major petroleum companies, the New York State Petroleum Council, and the American Petroleum Institute have cooperated with the Department of Health in surveying problems associated with waste oil. They have solicited the cooperation of personnel handling petroleum products at storage terminals and transportation facilities. Good housekeeping and conscientious operation in addition to protective and control devices can reduce the quantities of petroleum released or spilled in the state's waters.

Voluntary Action

The formation of local committees throughout the state has been encouraged. The Health Department provides technical assistance and serves in an advisory capacity. The Albany-Rensselaer Committee successfully promoted the purchase of a 1,500-ft (457-m) plastic boom that is stationed at a central point and available to area oil companies. The Health Department assists the responsible party in arranging for the removal and disposal of the oil once it is contained. In addition, an Alert Communications System has been established with state, interstate, and federal agencies; the province of Ontario is also a member. The system provides for immediate relay or exchange of information when spills or incidents occur that may have international or interstate implications.

In the Buffalo area, a joint municipal-federal research project has been initiated. The objectives are to: (*a*) develop techniques and equipment to remove oil from waters; and (*b*) identify type and source of oil releases.

State Regulation

Oil pollution control legislation passed during the current session of the legislature authorizes the Health Department to develop rules and regulations for petroleum and hazardous chemicals storage facilities.[2] Equipment and procedures required at storage facilities effective January 1, 1971, will be specified.

Federal Regulation

Federal oil pollution control legislation that became effective in April 1970 is mainly concerned with pinpointing the responsibility for large-scale disasters and specifying the size of penalty assessments for purposes of cleanup and reimbursement to those suffering damage from such catastrophes.

International Regulation

State and federal governments' intensified efforts to control oil pollution should reduce the number of incidents and damage. However, international control is mandatory. Ocean-going vessels ply international waters, fly foreign flags, and transport cargoes to and from many countries. The New York Harbor area is a base of operations for this type of activity. After delivering their cargoes of crude or processed petroleum, many tankers take on ballast water for smoother "riding" on the open seas. Before approaching their home ports to take on additional cargoes, the ships often discharge this oil-contaminated ballast water, causing large-scale slicks similar to those reported in Lake Ontario and off the metropolitan New York coastal area. These oil films can be formed many miles out to sea and gradually be carried toward shore by prevailing winds; the movement depends almost completely on wind and is only slightly affected by currents. The establishment of limits and areas for discharge of ballast waters in the open sea is of worldwide concern and should be internationally regulated. Other considerations are design and construction of the large ocean tankers to minimize cargo loss when mishaps occur.

Summary

Oil and other petroleum products represent major sources of pollution. The vast tonnage transported via pipelines, barges, trucks, and tankers in addition to storage must be rigidly controlled to minimize accidental spills and releases. State and national boundaries offer no protection when a spill occurs. New York waters can best be protected from the damaging effects of oil and other petroleum products by a cooperative program involving industry, consumers, and regulatory agencies. A step forward has been taken as a result of legislation adopted in 1970 by the federal and state governments and the continuing dialogue between official agencies and industry.

References

1. "'Torrey Canyon' Pollution and Marine Life." Cambridge Univ. Press, Cambridge, Eng. (1968).
2. "Public Law 91-224." New York State, 91st Congress, c 702, (Apr. 1970).

Additional Reference

Erickson, R. C., "Effects of Oil Pollution on Migratory Birds." In "Biological Problems in Water Pollutions." USPHS, U. S. Dept. of HEW, Washington, D. C. (1965).

OIL AND GAS WELLS—POTENTIAL POLLUTERS OF THE ENVIRONMENT?

A. Gene Collins

No detailed studies have been made of the ways in which drilling fluids, drilling muds, well cuttings, and well-treatment chemicals may contribute to pollution. Studies of well blowouts and possible development of communication between a freshwater aquifier and an oil-bearing sand have been made,[1] as have studies of possible pollution related to poor production practices.[2] The fact that the brines produced with oil and gas can contribute to pollution is well known,[3-6] but no universally satisfactory method for their disposal is available. Disposal of brine by solar evaporation in evaporating ponds has been investigated,[7] but final disposal of the residue salts needs further development. Some brines contain valuable minerals that are economically recoverable, and treatment or disposal of such brines should be coordinated with mineral recovery processes whenever possible.[8]

Several publications are available about oilfield brine disposal by subsurface injection into porous and permeable strata,[9-11] and the staff of the East Texas Salt Water Disposal Co. has prepared a comprehensive report[12] that describes gathering systems, pumps, treatment methods, and injection wells. Subsurface injection of oilfield wastes has proved to be a good method for disposal of potential water pollutants, but the results are not always satisfactory.[13,14] This disposal method has been blamed as the possible cause of earthquakes, and if a natural disaster such as an earthquake occurs, new faults or fractures in subsurface strata may provide communications between the strata containing the waste and the freshwater aquifer.[15-17]

Drilling

Drilling Fluids and Muds

The most modern drilling method is the rotary system, which requires circulation of drilling fluid for removal of drilled cuttings from the bottom of the hole to keep the drill bit and the bottom of the hole clean. The drilling fluids are pumped from ground surface through a drill pipe to the bottom of the hole and the bit and returned to the surface through the annulus outside the drill pipe. The flow of formation gas, oil, and brine into the drill hole is blocked by a fluid-mud column that produces a hydrostatic pressure that counterbalances or exceeds the formation pressures.

In certain geological environments, abnormally high fluid pressures are encountered, that is, the hydrostatic pressure is greater than 0.465 psi/ft of depth (0.106 bar/m of depth). When oil or gas wells are drilled into such an environment, there always is the possibility of a blowout unless elaborate precautions are taken and correct drilling muds are used. This situation can develop if degradation or sloughing off around the casing in a high-pressure zone occurs, allowing the pressured hydrocarbons to work their way along the outside of the casing to an upper zone (Figure 1).

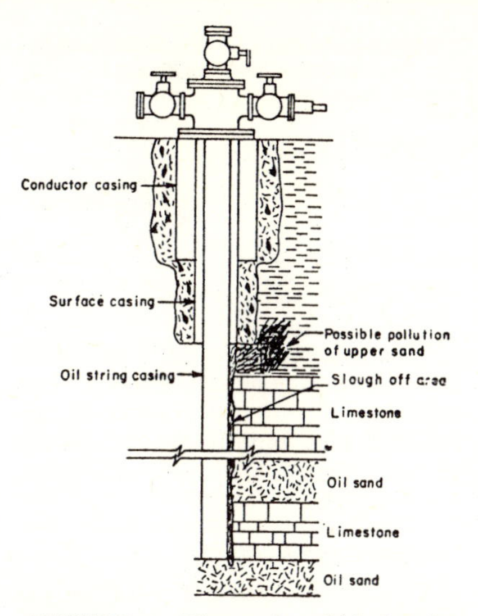

FIGURE 1.—Manner in which heaving shales or incompetent zones can slough off and allow a lower zone to communicate with an upper zone.

Drilling fluids may include gases, liquids, and solids suspended in liquids. Liquid drilling fluids include crude oil, fresh water, and salt water. Most of the solids suspended in liquids are called drilling muds and include suspensions of clays and other solids in water (water-base muds); suspension of solids in oil (oil-base muds); oil-in-water clay emulsions (oil-emulsion muds); and water-in-oil clay emulsion (inverted emulsion muds). Some of the compounds in drilling muds may be listed as follows[18,19]: quebracho extract; lignosulfonates, calcium, and chrome derivatives; acrylonitrites, such as hydrolyzed polyacrylonitrite; sodium salts of meta- and pyro-phosphoric acid; natural gums; tannins; molecularly dehydrated phosphates; subbituminous products; protocatechuic acid; barite; lignins, such as humic acids; bentonite; sugar cane fibers; lime; granular material, such as ground nutshells; corn starch; salt water; soluble caustic/lignin product; carboxy methyl cellulose; crude oil; sulfonated crude oil; oil emulsions; sodium chromate; anionic and nonionic surfactants; organophylic clay; soaps of long-chain fatty acids; phospholipids, such as lecithin; and asbestos.

Sulfonated drilling muds are prepared by: (*a*) sulfonating asphaltic crude oil with sulfuric acid, followed by neutralization with sodium silicate and ion exchanging with hydrated lime; or (*b*) absorbing concentrated sulfuric acid on a porous carrier (diatomaceous earth, for example) and then sulfonating asphaltic crude oil with acid carrier, followed by partial neutralization with sodium hydroxide and ion exchanging with hydrated lime.

The usual asphaltic crude oils that are used yield a 5- to 7-weight-percent carbon residue and have an American Petroleum Institute (API) gravity in the range of 26 to 31. Some blends may contain an 18° API asphaltic crude oil with a 12-weight-percent carbon residue blended with a paraffinic 42° API crude oil with a 0.5-weight-percent carbon residue. These muds are usually mixed with oil at the drilling site and used in the drilling operation. As the cuttings plus the used drilling mud are recovered from the well, the drilling mud usually is separated from the cuttings and reused. Some, of course, will be lost because it adheres to the cutting; therefore, some will present a possible water or land pollution hazard.[20]

The use of quebracho, starch, and carboxymethyl cellulose in formulating drilling muds has decreased in the past decade, whereas the use of chrome lignosulfonates has increased. The use of lime-treated mud systems has also decreased, whereas the use of low-solid muds, invert emulsions, and chrome lignosulfonate systems has increased.

Considerable money is invested in well muds, especially in the heavier muds; consequently, they are recovered for reuse. Such muds are used primarily for emergencies, such as lost circulation and high-pressure kicks from both gas and salt water. Many of the used muds are treated with high concentrations of lignosulfonates to produce a stable mud with excellent properties.

Possible sources of pollution from drilling fluids and muds are the fluids and muds that may be spilled during drilling, those that may escape into a subsurface freshwater aquifer, those that cling to the drill cuttings, and those that are not reused. The above list of constituents indicates that several constituents in drilling fluid and mud are capable of polluting water and land.

Chemical Treatment of Wells

Wells are treated with acids to increase the permeability of the reservoir rocks. This increases fluid flow and increases the recovery of oil and gas; it also improves fluid injection in secondary oil recovery and disposal operations. Hydrochloric, nitric, sulfuric, hydrofluoric, formic, and acetic acids are used. Such treatments produce soluble compounds including calcium chloride, sodium sulfate, sodium fluoride, and so forth, and, in addition, may leave partially spent acids in solution.

The volumes of acid used to acidize a well may vary from 500 to several hundred thousand gallons, depending on the amount of acid-soluble strata, the thickness of the horizon being treated, and the calculated productivity of the well.[21] Table I lists the approximate amounts of hydrochloric, formic, and acetic acids, used in the U. S. in 1 yr for oil- and gas-well treatment.

Other pollution problems can develop when the salt-enriched solutions plus any unspent acid are withdrawn from the well because subsequent disposal of these solutions is complicated by their tendency to form precipitates and their pH is low. It also is difficult to inhibit the acid-treatment solution to prevent corrosion,[22] and when corrosion does occur, the acid solutions and other fluids will escape at the point of pipe failure and pollute the adjacent zone, which may be a freshwater aquifer.

TABLE I.—Volume of Acids Used for Oil- and Gas-Well Treatment

Acid	Amount Used (gal/yr)
Hydrochloric	8.7×10^7
Formic	2.0×10^5
Acetic	1.0×10^5

Note: Gal × 3.785 = l.

Corrosion Inhibitors

A universal "super" inhibitor has evaded the researcher for 40 yr.[21] Such an inhibitor would be useful to prevent steel casing and tubing from corroding as a result of acid treatment of a well. The best available high-temperature inhibitor is a combination of sodium arsenite with an alkyl phenolethylene oxide surfactant, and arsenic-type inhibitors have been used for both low- and high-temperature applications since the 1930's. Table II lists some of the inhibitors used in the U. S.[23]

Other Additives

To reduce friction, reduce loss, sustain permeability, prevent emulsions from forming, and prevent precipitation, additives are added to the oil- or gas-well systems. Table III lists some of the compounds used for these pur-

TABLE II.—Types and Amounts of Inhibitors Used in Oil- and Gas-Well Treatment

Inhibitor	Amount Used (lb/yr)
Sodium arsenite	1.0×10^6
Imidazoline	1.25×10^6
Abiethylamine	7.0×10^5
Coal tar derivatives	2.5×10^5
Acetylenic alcohol-alkylpyridine	3.0×10^5

Note: Lb/yr × 0.454 = kg/yr.

TABLE III.—Types and Amounts of Other Additives Used in Oil- and Gas-Well Treatment

Additive	Amount Used (lb/yr)
Lactic acid (44%)	5.75×10^5
Citric acid	2.0×10^4
Alkylaryl sulfonic acid	5.0×10^5
Zirconium oxychloride (20%)	2.5×10^5
Quatenary ammonium derivatives	2.0×10^5
Polymers	1.0×10^5
Polyacrylamide	6.0×10^5
Guar gum	5.75×10^6
Fluid loss agents	1.8×10^6
Emulsion preventers	4.5×10^5

Note: Lb $\times$ 0.454 = kg.

poses and the approximate amounts used in 1 yr.

Possible Pollution from Petroleum

An opening or hole from the ground surface to a subsurface oil- or gas-bearing formation is a well. Such an opening usually is lined with a metal pipe cemented in place, and production equipment is fastened to the cased hole to regulate and control oil or gas withdrawal rates. Before a well is drilled, some knowledge of the geologic formations to be penetrated is useful, as is knowledge of the approximate depth of the target petroleum-bearing zone. This information is needed so that the appropriate diameter, length, and type of oil-string casing can be selected in planning the well.

Most states have laws requiring the setting of surface casing to protect the freshwater subsurface sands from invasion by brines and hydrocarbons from deeper horizons. Therefore, a minimum of two strings of casing—the surface casing and the oil-string casing—will almost always be required. Additional strings of casing may be required if heaving shales are found while drilling progresses if abnormal pressures are encountered or if a zone of lost circulation is found. Each additional string of casing requires more capital and increases the cost of the well.

If appropriate precautions are not taken in planning, drilling, and completing an oil or gas well, disastrous consequences can occur. For example, during drilling operations or pulling of the drill pipe, a well may blow out if adequate mud pressure is not maintained. Such a situation may develop if the mud line is accidentally broken or if the well casing is not properly cemented to competent zones. Figure 2 illustrates what might occur if fluid from a high-pressure well escapes into an incompetent zone and develops communication of a lower hydrocarbon-bearing horizon with an upper sand.

Production

Possible Pollution from Petroleum

Crude oil in excessive amounts is detrimental to vegetation; oily wastes on surface waters can cause a fire haz-

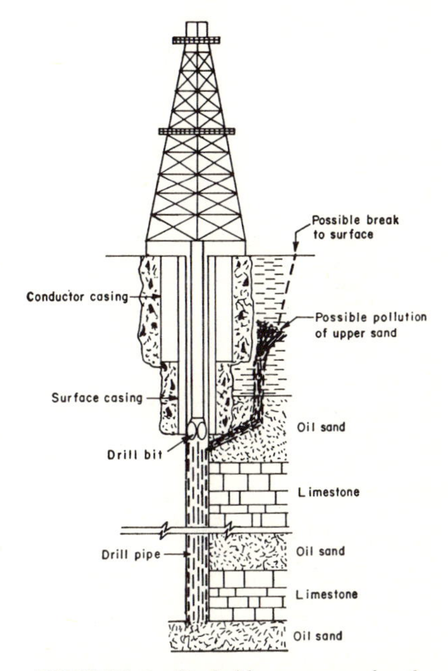

FIGURE 2.—Probable manner whereby a well blowout can develop communication between an upper sand and a lower sand.

ard, can be deleterious to fish life, and gradually will combine with particulate matter, sink, and thus pollute the bottom of the stream or lake. Further, crude oil has destructive effects on fowl that may swim in the polluted water and may damage the surrounding flora and the surrounding beaches. Mercury concentrations in excess of 20 mg/l are present in some crude oils. In essence, then, it can be assumed that any excessive amount of produced crude oil that finds its way to surface lands or waters will cause pollution.

The composition of the crude oil that pollutes the water or land will determine the extent and type of pollution. For example, some heavy crude oils possess a specific gravity of about 1, contain about 5-weight-percent sulfur, and have an overall minimum boiling point of about 270°C. Conversely, some light crude oils contain virtually no elements other than carbon and hydrogen, have 0.8 or less specific gravity, and distill below 270°C. The major nonhydrocarbons in crude oils are basic and nonbasic nitrogen and sulfur compounds and acidic and nonacidic oxygen compounds. Usually the nonhydrocarbons are more highly concentrated in the heavier portions of the crude oils. In an overall classification, most crude oils can be classified as naphthenic, paraffinic, or intermediate; the naphthenic type usually is the heaviest, the paraffinic the lightest.

Once the crude oils escape on land or water, they are subjected to evaporation, oxidation, solution, dispersion, and utilization by microorganisms. The lighter crude oils will evaporate more readily than will the heavy ones. The lower hydrocarbons, (methane and benzene, for example) though relatively insoluble in water, will be more soluble than the higher-molecular-weight hydrocarbons; the crude oils containing sulfur compounds probably will oxidize less rapidly than will those containing metallo compounds. Crude oils, when spread on salt water, such as the sea, will quite rapidly form highly stable water-in-oil emulsions, as was exhibited in the *Torrey Canyon* disaster.

This type of emulsion forms thick blobs of oil that are fairly resistant to dispersal, oxidation, and bacterial reactions. The reason that this type of emulsion forms with salt water has not been clearly established. A means of readily reverting such emulsions to an oil-in-water type would be desirable for quick dispersal.[24]

Emulsions of petroleum and brine or mixtures of crude oils and sand that are difficult to break can be found on surface disposal ponds. Should these ponds overflow, the surrounding land or surface streams will be polluted. Crude oil also may escape from leaky connections, from improperly plugged wells, from improperly cased and cemented wells, from holes in lines or storage tanks, or as a result of an accident. Burning of the petroleum or emulsions, or both, that enter brine ponds can contribute to air pollution, and all of the petroleum will not be completely destroyed by the fire.

Oil production may produce pollution in onshore or offshore areas from blowouts of the wells, dumping of oil-based drilling muds and oil-soaked cuttings, or losses of oil or brine in production, storage, and transportation. Over 200,000 miles (320,000 km) of pipeline operating at pressures to 1,000 psi (70.3 kg/sq cm) are used throughout the country and in offshore areas. Any rupture or accidental puncture of any of these lines results in pollution.

Possible Pollution from Natural Gas

Blowouts of natural gas wells will contribute to pollution, especially if the natural gas contains appreciable quantities of hydrogen sulfide. Many gas wells contain enough hydrogen sulfide to pollute any fresh water they may contact. Such contact may develop if a well is faulty and communication between the gas zone and an upper freshwater zone occurs. Brines associated with hydrogen-sulfide-bearing gas zones also will contain appreciable quantities of the sulfide.

Possible Pollution from Oilfield Brines

Waters associated with petroleum in subsurface formations usually contain many dissolved ions. Those most commonly present in greater than trace amounts are sodium (Na^+), calcium (Ca^{2+}), magnesium (Mg^{2+}), potassium (K^+), barium (Ba^{2+}), strontium (Sr^{2+}), ferrous iron (Fe^{2+}), ferric iron (Fe^{3+}), chloride (Cl^-), sulfate (SO_4^{2-}), sulfide (S^{2-}), bromide (Br^{2-}), bicarbonate (HCO_3^-), and dissolved gases such as carbon dioxide (CO_2), hydrogen sulfide (H_2S), and methane (CH_4). The stability of petroleum-associated brine is related to the constituents dissolved in it, the chemical composition of the surrounding rocks and minerals, the temperature, the pressure, and the composition of any gases in contact with the brine.[25]

Scale inhibitors are added to waters and brines to prevent the precipitation reactions. Some of the chemicals used in these inhibitors are ethylenediamine tetraacetic acid salts, nitrilotriacetic acid salts, sodium hexametaphosphate, sodium tripolyphosphate, and sodium carboxymethyl cellulose.

Knowledge of the oxidation state of dissolved iron in brines is important in compatibility studies. Brines in contact with the air will dissolve oxygen, and their Eh generally will be from 0.35 to 0.50. Brines in contact with petroleum in the formation normally will have an Eh lower than 0.35, as will waters in contact with reducible hydrocarbons.[26] Any change in the oxidation state of brine containing dissolved iron may result in the deposition of dissolved iron compounds.

The sediments or precipitate formed from brines can cause environmental pollution directly or indirectly. For example, if the produced brines are stored in a pond, the sediments may cause soil pollution; if the brines are injected into a disposal well, the sediments may plug the face of the disposal formation, resulting in the necessity for higher injection pressures, which may rupture the input system.

The amount of salt water or brine produced from oil wells varies considerably with different wells and depends on the producing formation and the location, construction, and age of the well. Some oil wells produce little or no brine at first, but as time goes by, they gradually produce more and more brine. As some wells become older, the produced fluids may be more than 95 percent brine, that is, for each barrel of oil coming to the surface, 100 bbl or more of brine also is produced. The produced brines differ in concentration but usually consist primarily of sodium chloride in concentrations ranging from 5,000 to more than 200,000 mg/l; the average probably is about 40,000 mg/l. For comparison, note that seawater contains about 20,000 mg/l of chloride. One barrel of brine containing 100,000 mg/l of chloride will raise the chloride content of 400 bbl or about 17,000 gal (64,000 l) of fresh water above the maximum recommended for drinking water. Petroleum-associated brines may escape and contact fresh water or soil in different ways. For example, to protect the upper fresh waters from the deeper mineralized waters that might rise in the drilled well, the upper portion of the well is sealed by a string of cemented surface casing. If a well has insufficient surface casing, an avenue may be provided for the escape of brines if they are under sufficient hydrostatic head to cause them to rise in the hole to the surface or to the level of freshwater sands.

Handling the tremendous volume of brine produced simultaneously with petroleum is hazardous. Basically, the problem is to handle and dispose of the brine in such a manner that it does not contact soil or fresh water and cause detrimental pollution.

Currently, some produced brines are being discharged into approved surface ponds, whereas most brines are returned underground for disposal or to repressure secondary oil or gas recovery wells. The discharge of brines to any surface drainage is strictly prohibited in most states. Potential water and soil pollution problems are associated with both disposal methods. For example, if the surface pond is faulty, the brine will contact the soil, and various chemical reactions will occur between the soil and the brine. Sometimes the brine will pass through the soil, reappear at the surface, and produce scar areas; sometimes it will pollute the soil, and leaching will pollute surface streams or shallow subsurface aquifers.

Residual Salt Concentrations Beneath or Near Abandoned Unsealed Disposal Ponds

Unsealed surface ponds used for the disposal of oilfield brines have polluted fresh surface waters, potable groundwaters, and fertile land. Because of chemical and physical phenomena and dispersion, the movement of soluble pollutants from these ponds is complex. For example, the soluble pollutants move slowly in relation to the soil water flow rate, and dispersion effects a displacement which causes the contaminated zone to grow.

The Kansas State Department of Health studied the soils beneath and near an old unsealed brine disposal pond that had been abandoned for 10 yr. During its use, the pond received more than 32,000 tons of salts, and most of those soluble salts probably escaped by soil leaching and downdrainage and penetrated below the underlying limestone formation. Eleven test holes were drilled into the soil and shale beneath and adjacent to the pond, both above and below the natural drainage slope. Chemical analysis of the test hole core samples indicated that more than 430 tons (about 1.4 percent of the original) of soluble residual salt still remained to be leached out of the soil and shale in the pond area. This amount of soluble or leachable salt remaining in the area indicates that the return of the subsurface water and soil to their prepollution level is a very slow process and may take several decades. Network pollution zones seem to form where formation fracture conjugates occur. Leaching seems to be entirely dependent on the flushing mechanism provided by meteoric water.

The cation concentrations in the clay minerals were evaluated by x-ray diffraction techniques to trace cation transportation rates. Chloride analysis was selected as the most useful single means of detecting the presence of oilfield brine pollution, but the associated cation concentration should also be determined to formulate a more complete picture. Cation adsorption studies are apparently useful in differentiating brine-polluted soil and shale, clay mineral studies provide the information on the environmental characteristics of the pollution media, and cation exchange information aids in explaining the apparent differential transportation rates of ions in brine seepage solutions.[27, 28]

Disposal

Subsurface Disposal

A problem associated with subsurface brine disposal is casing leaks in the disposal well, which could allow the brine to enter freshwater aquifers. Figure 3 shows how an improperly designed disposal well and a leaky oil well can pollute a freshwater aquifer. Erroneous geologic information about the subsurface formation into which the brine is being pumped presents another problem. Brine usually is pumped into a subsurface formation

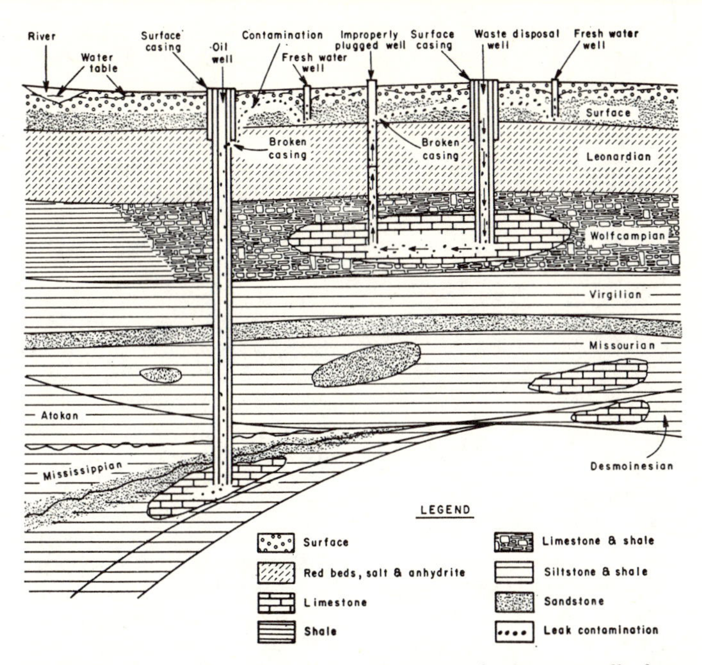

FIGURE 3.—Route by which salt water enters freshwater wells from faulty oil or disposal wells.

that contains similar brine; however, exact knowledge of the faulting and fracturing of such a subsurface formation is difficult to obtain. Because the brine is pumped into the formation, bottomhole pressure must not exceed 1 psi/ft (0.23 kg/sq cm/m) of overburden, or the hydraulic pressure may cause fracturing, and, in time, the wastes may migrate to a freshwater zone.

Petroleum-associated brines from two different formations may form precipitates if they are mixed. For example, with a well used for disposal of brines from several producing oil wells, it is imperative that precautions be taken in mixing and treating the brines before injection. If the brines are incompatible and inappropriate precautions are taken, there is a possibility that deposits will form and filter out on the face of the injection formation, thus reducing the permeability. The quantity of deposits formed from incompatible brines depends on ions present. The more common deposits resulting from reactions of incompatible brines are gypsum ($CaSO_4 \cdot 2\ H_2O$), anhydrite ($CaSO_4$), aragonite ($CaCO_3$), calcite ($CaCO_3$), celestite ($SrSO_4$), barite $BaSO_4$), troilite (FeS), and siderite ($FeCO_3$).

Subsurface brine disposal can be categorized as confinement or containment; confinement is the placement of brines in a horizon where any movement can be controlled or monitored, while containment is the placement that precludes the movement of the brines out of a formation or zone. Note that containment cannot be used for an unlimited supply of brine, but that confinement necessitates the monitoring of the migration of the brines. The knowledge necessary to define the hydrodynamics of brines injected into subsurface environments is expensive to obtain, and much of the necessary fundamental knowledge of subsurface formations is not available. Formations into which brines are often pumped for disposal are called salaquifers, and these zones consist of permeable sedimentary rock. Some information needed before such a zone can be used for disposal operations is as follows:

1. The size of the zone;
2. The migration of the brine in the zone (whether it might reappear in another zone or perhaps migrate to the surface);
3. The mechanisms that control movement in a given salaquifer or out of it;
4. The steps necessary to assure containment or confinement of the brine within the salaquifer.

It is difficult, if not impossible, to develop adequate knowledge concerning how or where escape channels may occur from a salaquifer. Test drilling is the only known method that can provide such knowledge, and the drilling and subsequent evaluation are expensive.[29]

Joint ownership of disposal systems by several companies helps minimize installation and maintenance costs. Brines can be gathered through common lines and accumulated at a central location so that one disposal well serves many producing wells. Investment costs depend primarily on the brine characteristics from the producing formations, the receptivity of the disposal formation, and the condition of the surface soil for gathering-line installation. Where the brine production is relatively small, a complete system can be installed for less than $500/producing well, but if large volumes of brine are produced, the disposal well may be able to service only a few producing wells, and the cost may be $8,000/well or more. In 1968 the operating costs in representative fields in the Permian Basin amounted to 6.7 mil/bbl for 38 mil bbl of brines.[11, 30]

In 1967 in Texas there were 41,000 active oil wells and about 6,900 active gas wells, and more than 5 mil bbl/day of brine was produced. That amount of brine contains approximately 65,000 tons (58,500 metric tons) of salts and that amount of daily produced salt can pollute 26 bil gal (98 bil l) of fresh water to the point at which it would not be acceptable as drinking water.

Water flooding of oil sands was begun in Bradford Field, Pa., in 1907, and was developed into a systematic operation after 1934. Considerable care must be exercised in using this method to recover oil because there is a danger that the reinjected brine will migrate to freshwater streams.

Subsurface disposal of oilfield brines, as well as of industrial wastes, is being increasingly used to replace surface disposal.[31, 32] The ideal conditions for formations used for such disposal are large areal extent, high permeability and porosity, overlying and underlying aquicludes, low internal pressure, salaquifer, compatible fluids, no unplugged wells open to an outcrop, and uniformity. The reservoir used for disposal must be very large, but even so, the amount of fluid that can be injected is limited.

Many things are not known about what happens within a formation used for disposing of wastes. For example, many wastes are low in pH, and apparently no studies have been made of how the pH changes with time within the formation. Conceivably the acid can react with the rock and perhaps break out. It is known that most accidental fractures of the formation or the overlying aquiclude will be horizontal if the well is no deeper than 1,000 ft (300 m). However, if the disposal well is deeper than 1,500 ft (450 m), the fracture orientation is likely to be vertical, and vertical fractures can, if large enough, cause communication with an upper zone.

Salt water under pressure will attempt to escape from any type of confinement. The salt water may escape through fractures because of a mechanical failure within the individual well system, through an old drill hole that penetrates the injection zone, or through a natural fault system caused by a recent earthquake.

Recovery of Valuable Elements before Disposal

Elements found in some brines in economic concentrations are magnesium, calcium, potassium, lithium, boron, bromine, and iodine. Many of them are recovered by chemical companies from seawater, salt lakes, and subsurface saline waters.[33, 34]

Factors that must be considered in evaluating a saline water as an economic resource are the cost of bringing it to the factory, the cost of the recovery process, and the cost of transporting the recovered products to market. When a brine is produced only for the purpose of recovering its dissolved chemicals, a prime factor is the cost of pumping the brine. It will cost less to produce the brine from a shallow well than from a deep well. Therefore, disregarding other factors, a brine must not only contain a certain amount of recoverable chemicals, but it must be available in large quantity before it can be considered economically valuable, and the farther it must be pumped, the more chemicals it must contain. Today the possibility of recovering elements from brines that are pumped to the surface is increasingly important because the brines present a pollution hazard if their disposal is improper.

Conclusions and Recommendations

1. The amount of pollution attributable to drilling fluids, drilling muds, well-treatment chemicals, and drill cuttings needs evaluation. Improved methods are needed to prevent and abate pollution from oilfield brines, spilled petroleum, and petroleum-related fluids and gas that are inadvertently allowed to enter freshwater areas. Permanent disposal techniques for undesirable petroleum-related materials are needed.

2. Geochemical research is needed to determine what happens within a salaquifer used for disposal and what happens to related salaquifers. Hydrogeological studies of the salaquifers used for disposal are needed. Disposal into salaquifers is limited and cannot continue indefinitely; therefore, other methods of disposal need development. Studies concerning rates of reaction of petroleum-related materials with soils and water must be conducted to permit understanding and development of appropriate environmental protection techniques.

3. Sedimentary basins contain saline waters that are enriched in valuable elements. Chemical companies are recovering minerals from some of these waters and will eventually exploit others for their mineral wealth. The cost of producing these petroleum-associated brines is covered because they are brought to the surface with petroleum. Their mineral wealth should be extracted before disposal to help pay for the disposal, to prevent pollution, and to conserve valuable natural resources.

References

1. Vedder, J. G., *et al.*, "Geology, Petroleum Development, and Seismicity of the Santa Barbara Channel Region, Calif." U. S. Geol. Surv. Prof. Paper 679 (1969).
2. Schmidt, L., and Wilhelm, C. J., "Disposal of Petroleum Wastes on Oil-Producing Properties." U. S. Bur. Mines Rept. of Inv. 3394 (1938).
3. Crouch, R. L., "Investigation of Alleged Ground-Water Contamination Tri-Rue and Ride Oil Fields, Scurry County, Texas." Texas Water Comm., LD-0464-MR (1964).
4. Grandone, P., and Schmidt, L., "Survey of Subsurface Brine-Disposal Systems in Western Kansas Oil Fields." U. S. Bur. Mines Rept. of Inv. 3719, (1943).
5. Taylor, S. S., *et al.*, "Study of Brine-Disposal Systems in Illinois Oil Fields." U. S. Bur. Mines Rept. of Inv. 3534 (1940).

6. Wilhelm, C. J., and Schmidt, L., ''Preliminary Report on the Disposal of Oil-Field Brines in the Ritz-Canton Oil Field, McPherson County, Kansas.'' U. S. Bur. Mines Rept. of Inv. 3297 (1935).
7. Gunaji, N. N., and Keyes, C. G., Jr., ''Disposal of Brine by Solar Evaporation.'' Office of Saline Water Res. and Develop. Progress Rept. No. 351 (1968).
8. Collins, A. G., ''Here's How Producers Can Turn Brine Disposal into Profit.'' *Oil & Gas Jour.*, **64**, 27, 112 (1966).
9. Morris, W. S., ''Salt Water Disposal from the Engineering Viewpoint.'' Presented to the Research Committee of the Interstate Oil Compact Commission, Dallas, Tex., May 31, 1956.
10. Payne, R. D., ''Salt Water Pollution Problems in Texas.'' *Jour. Petrol. Technol.*, **18**, 1401 (1966).
11. Rice, I. M., ''Salt Water Disposal in the Permian Basin.'' *Producers Monthly*, **32**, 3, 28 (1968).
12. East Texas Salt Water Disposal Company, ''Salt Water Disposal East Texas Field.'' Petroleum Extension Service, Austin, Tex. (1953).
13. Donaldson, E. C., ''Subsurface Disposal of Industrial Wastes in the United States.'' U. S. Bur. Mines Inf. Circ. 8212 (1964).
14. Watkins, J. W., *et al.*, ''Feasibility of Radioactive Waste Disposal in Shallow Sedimentary Formations.'' *Nucl. Sci. & Eng.*, **7**, 133 (1960).
15. Bardwell, G. E., ''Some Statistical Features of the Relationship Between Rock Mountain Arsenal Waste Disposal and Frequency of Earthquakes.'' *Mountain Geol.*, **3**, 23 (1966).
16. Evans, D. M., ''The Denver Area Earthquakes and the Rocky Mountain Arsenal Disposal Well.'' *Mountain Geol.*, **3**, 23 (1966).
17. Warner, D. L., ''Subsurface Injection of Liquid Wastes.'' In ''Natural Gas, Coal, Ground Water Exploring, New Methods, and Techniques in Resources Research.'' Univ. of Colorado Press, Boulder, 107 (1966).
18. Caraway, W. H., ''Quebracho in Oilwell Drilling Fluids.'' *Petrol. Engr.*, **25**, 12, B-81-92 (1953).
19. Simpson, J. P., *et al.*, ''Some Recent Advances in Oil Mud Technology.'' SPE-150, Presented at 36th Annual Meeting of Society of Petroleum Engineers of AIME, Dallas, Tex., Oct. 8–11, 1961.
20. Messenger, J. U., ''Composition, Properties and Field Performance of a Sulfonated Full Oil Phase Mud.'' *Jour. Petrol. Technol.*, **15**, 259 (1963).
21. Hurst, R. E., ''Market for Completion and Stimulation Chemicals.'' Presented before the Div. meeting of Petrol. Chem., Inc., Amer. Chem. Soc. meeting, Houston, Tex., Feb. 22–27, 1970.
22. Harris, O. E., *et al.*, ''High Concentration Hydrochloric Acid Aids Stimulation Results in Carbonate Formations.'' *Jour. Petrol. Technol.*, **18**, 1291 (1966).
23. Cowan, J. C., ''Some Secondary Properties of Chemicals Used for Mineral Scale Inhibition.'' Presented before Div. of Petrol. Chem., Inc., Amer. Chem. Soc. meeting, Houston, Tex., Feb. 22–27, 1970.
24. Dean, R. A., ''The Chemistry of Crude Oils in Relation to Their Spillage on the Sea. The Biological Effects of Oil Pollution on Littoral Communities Supplement to Field Studies.'' Vol. 2, Field Studies Council (1968).
25. Fulford, R. S., ''Effects of Brine Concentration and Pressure Drop on Gypsum Scaling in Oil Wells.'' *Jour. Petrol. Technol.*, **20**, 559 (1968).
26. Hem, J. D., ''Stability Field Diagrams in Iron Chemistry Studies.'' *Jour. Amer. Water Works Assn.*, **53**, 211 (1961).
27. Bryson, W. R., *et al.*, ''Residual Salt Study of Brine Affected Soil and Shale Potwin Area—Butler County, Kansas.'' Bull. 3-1, Kansas State Dept. of Health, Topeka (1966).
28. Siever, R., ''Establishment of Equilibrium Between Clays and Sea Water.'' *Earth & Plan. Sci. Letters*, **5**, 106 (1968).
29. Drescher, W. J., ''Hydrology of Deep-Well Disposal of Radioactive Liquid Wastes.'' In ''Fluids in Subsurface Environments.'' A. Young and J. E. Galley [Eds.]. Memoir 4, 399, Amer. Assn. Petrol. Geol. (1965).
30. Research Committee, Interstate Oil Compact Commission, ''Subsurface Disposal of Industrial Wastes.'' Interstate Oil Compact Commission, Oklahoma City, Okla. (1968).
31. Enright, R. J., ''Oilfield Pollution.'' *Oil & Gas Jour.*, **61**, 76 (June 24, 1963).
32. Research Committee, Interstate Oil Compact Commission, ''Production and Disposal of Oilfield Brines in the United States and Canada.'' Interstate Oil Compact Commission, Oklahoma City, Okla. (1960).
33. Brennen, P. J., ''Nevada Brine Supports a Big New Lithium Plant.'' *Chem. Eng.*, **73**, 17, 86 (1966).
34. Collins, A. G., ''Finding Profits in Oil-Well Waste Waters.'' *Chem. Eng.*, **77**, 20. 165 (1970).

PROBLEMS IN WATER ANALYSIS FOR PESTICIDE RESIDUES

A. BEVENUE, T. W. KELLEY AND J. W. HYLIN

INTRODUCTION

The examination of waters for pesticides has included samples of potable waters, fresh water streams, lakes, rivers, the oceans, and even sewage outfall areas. Without doubt, there will be non-pesticide chemicals in some samples that will possess analytical characteristics similar to some pesticides when they are examined by electron capture gas chromatography. In addition, false data may be acquired from extraneous sources during the manipulation of the samples in the analytical laboratory which will not be eliminated by confirmatory techniques, such as thin-layer chromatography (TLC), unless certain precautionary measures are taken prior to the analysis of the samples.

This report reviews some of these problems and also brings to the reader's attention several areas of analytical interferences which, to our knowledge, have not been specifically mentioned heretofore in the literature. The experienced analyst may be aware of these problems. However, with the increased interest in environmental studies coupled with the required establishment of many new laboratories possibly staffed with personnel inexperienced in trace analysis techniques, this report may aid the analyst in avoiding some unforeseen problems in the analysis of waters for pesticides.

When large samples of water (five or more gallons) are extracted for analysis and the extract is concentrated to a small volume, the suspected pesticide(s) in the water may be confirmed by TLC and spray reagent techniques if the pesticide residue in the extract is in the microgram range. However, the trend toward the use of small grab samples (one gallon or less) and the desire to find and report residues in the part per trillion, or fraction thereof, range eliminates the possibility of using the spray technique for verification, because of the detectability limits of the stain reagent. Under the latter conditions, the area of the developed TLC plate containing the suspected pesticide is eluted with a suitable organic solvent and the concentrated extract therefrom is again subjected to gas chromatographic analysis. Extraneous interferences are magnified on the recorder chart unless (1) special precautions are taken with the organic solvents, the glassware, and other equipment used in the analytical procedure and (2) the thin-layer adsorbent is completely free from organic contaminants.

EXPERIMENTAL

Organic solvents

Organic solvents of "reagent grade" quality cannot be used for pesticide residue analysis in the nanogram–picogram range, because of contaminants in the solvent which will be magnified on the gas chromatograph recorder chart when the final concentrated extract is applied to the gas chromatograph. It is inexcusable to use such reagents in the analytical procedure since high-purity solvents are now commercially available. If necessary, reagent-grade solvents should be redistilled in an all-glass system; however, the redistilled reagent should be checked before use.

Glassware and other equipment used prior to TLC and GC analysis of the water sample

LAMAR *et al.*[1] recommended heating all glassware, except volumetric ware, overnight at 300° prior to use with water samples. The volumetric glassware was cleaned with a solution of sodium dichromate in concentrated sulfuric acid. They also warned against the use of rubber, cork, or plastic stoppers for water sample containers. The Federal Water Pollution Control Administration water analysis manual[2] recommended heating the glassware at 400° if the type and size of glassware permitted such drastic treatment. AMOS[3] reported variable results—some satisfactory, some poor—when glassware was soaked in acid or base solutions or detergent solutions.

Plastic tubing used in vacuum equipment to remove sections of the silica powder from TLC plates have contributed organic contaminants to the powder[3,4].

Soxhlet extraction thimbles (Whatman cellulose) contain substances which will produce pseudo-pesticide peaks on the gas chromatogram unless the thimbles are solvent-extracted prior to use[3,5].

The following additional precautionary measures are suggested based on studies conducted in our laboratory. The glass jars used for developing the TLC plates may not tolerate the stress of heat treatment; therefore, sodium dichromate–sulfuric acid solution should be applied to the interior walls of the jar, followed by rinsing with water, acetone, and hexane. Whatman filter paper sheets are commonly used as liners in the chromatographic tank to saturate the interior of the tank with the vapors of the developing solvent. This practice cannot be tolerated in water analysis confirmatory work, because the paper may contaminate the developing solvent with organic materials which will be transferred to the TLC silica gel plate and finally to the concentrated eluted extract. Although the separation of certain groups of chlorinated pesticides may not be as efficient without the use of the paper liner, the eluted fractions from the TLC plates will give satisfactory results on the gas chromatograph for confirmatory analyses.

The syringes used for gas chromatograph samplings must be scrupulously clean and may require copious sequential washes with alcohol, acetone, and hexane accomplished by passing the solvents through the barrel of the syringe with the aid of a vacuum pump or water aspirator apparatus.

The inclusion of any glassware which contains ground-glass sections, such as glass-stoppered centrifuge tubes or volumetric flasks, will add to the analytical problem. Heating the glassware will not remove the contaminants from the ground, glass sections. Lengthy periods of washing with copious amounts of solvents may clean the ground-glass areas, but the procedure is impractical. If this type of glassware must be used, the contents of the container should not be poured from one container to another; the transfer should be made preferably with clean, heat-treated, disposable glass pipettes.

Four cleaning procedures for glassware were examined, each of which consisted of the same initial preparation as follows: Silica gel (0.1 ml dry volume), known to contain organic contaminants, was added to each of a series of centrifuge tubes (Kontes No. 410550, 5 ml capacity) and also to each of a series of Chromaflex sample tubes (Kontes No. 422560, 2 ml capacity); only 0.01 ml dry volume of gel was added to each sample tube. Hexane (0.5 ml) was added to each tube; the contents were coated on the interior walls of the tube by means of agitation on a vortex mixer. The contents of each tube was discarded. With preknowledge of the organic contaminant content of the silica gel, the above-mentioned amounts of gel were added to each tube to approximate the amount of gel that would be scraped from a TLC plate for further study and which would also approximate the amount of background contamination observed on the gas chromatograph recorder chart. Each tube was then washed with tap water and a nylon-bristle brush to remove adhering particles of gel from the walls of the tubes. Each tube was then rinsed with copious amounts of distilled water. The four subsequent cleaning procedures with sets of the above-mentioned contaminated tubes are outlined in Table I. Auxiliary glassware used throughout the analytical procedure was cleaned in a similar manner.

Subsequent to the cleaning procedures outlined in Table I, 0.5 ml of redistilled

TABLE I

RINSING SOLUTIONS IN GLASSWARE CLEANING PROCEDURES

Method			
1	*2*	*3*	*4*
Ethanol[a]	Dichromate–H_2SO_4[b]	Acetone	Dichromate–H_2SO_4[b]
Acetone	Tap water		Tap water
Hexane	Distilled water		Distilled water
	Acetone		Acetone
Air-dry	Air-dry	Air-dry	Air-dry
		Heat[c]	Heat[c]

[a] Glassware rinsed three times with each solvent in order of the listed sequence.
[b] Glass was soaked for 16 h in a solution of sodium dichromate–concentrated sulfuric acid.
[c] Glass was heated in an air oven for 16 h at 200°.

hexane was added to each centrifuge tube by means of a heat-treated disposable pipette, to avoid contacting the ground glass area of the tube with the solvent. The tube was then agitated for about 30 sec on a vortex mixer to wash the wall of the tube with the solvent, again being careful not to wet the ground glass area of the tube. The hexane was transferred, by pipette, from the centrifuge tube to the Chromaflex sample tube. The procedure was repeated four times, combining all hexane fractions in the sample tube. The hexane content of the sample tube was concentrated to about 0.02 ml with the aid of a stream of filtered nitrogen, and aliquots of the solution were applied to the gas chromatograph to check the efficiency of the glass cleaning procedure.

Referring to Table I, method No. 1 will not completely remove contaminants from the glassware. Methods 2, 3, and 4 will remove all contaminants and any one of the three procedures can be recommended. However, because of its relative simplicity, method No. 3 is preferred. It is apparent that organic solvents alone will not remove firmly bonded organic contaminants from glass; the more drastic treatment with an oxidizing reagent and a concentrated mineral acid and/or heat are prerequisites for contaminant-free glass equipment.

SCHAFER *et al.*[6] studies on pesticides in "drinking" waters described a stirring bar mechanism to mix thoroughly hexane and the water sample in gallon jugs for the extraction of the pesticides. Using this technique, it has been our experience that if Teflon magnetic stirring bars are used, the water sample will be grossly contaminated if the Teflon bars have been in previous contact with plant extracts or other biological media. If one contemplates using this mixing technique in water analyses, only new Teflon bars should be used and they should be screened for possible contamination properties prior to use.

Silica gel adsorbents

The transition in the past 10–15 years from the milligram range to the nanogram–picogram range in chemical analysis techniques must be considered in the following discussion on silica gel adsorbents. MILLER AND KIRCHNER[7] noted that silicic acid adsorbents contained as much as 100 mg of a yellow oily material in 100 g of adsorbent which was soluble in ethyl acetate or acetone and which, if present, interfered with UV and fluorescein tests on the chromatograms. STANLEY *et al.*[8] prewashed silica gel TLC plates with petroleum ether, followed by a continuous wash for 4 to 16 h with ethyl alcohol, to remove organic materials that would interfere with diphenyl analysis in citrus fruits. This type of cleanup for the silica gel was apparent-sufficient for the measurement of milligram quantities of diphenyl by a spectrophotometric procedure. BOWYER *et al.*[9] extracted silicic acid with a chloroform–methanol (2:1) mixture to remove lipid contaminants prior to using the silicic acid for the analysis of fatty acids, also in the milligram range. BROWN AND BENJAMIN[10] noted that organic contaminants in commercially available silica gels obscured the desired spots on the acid-sprayed chromatogram and recommended washing the plates with a mixture of diethyl ether–methanol (20:80). AMOS[3] extracted various grades of silica gel with acetone and obtained residues of dark brown oils in amounts ranging from 1.0 to 39 mg per 100 g of gel. KOVACS[11] washed silica gel plates with distilled water prior to use, to remove "chlorides" that would interfere with the $AgNO_3$ spray reagent subsequently used for the detection of pesticides at the 0.10 μg

level. Later, KOVACS[12] discontinued the use of silica gel for pesticide residue analysis and replaced the gel with the adsorbent aluminum oxide, because of the high levels of "chlorinated" impurities in the silica gel. SMITH AND EICHELBERGER[13] used silica gel, as purchased, for the separation of pesticides extracted from a water sample; the amount of each pesticide applied to the TLC plates was about 0.2 mg. Sections of the developed silica gel plates were eluted for gas chromatography confirmatory analysis. Although the suspected pesticides were "confirmed" by this procedure, other unknown components sensitive to the electron capture detector were noted on the gas chromatogram. GEISS *et al.*[4] noted that silica gel contained organic contaminants, and the problem was aggravated by the use of plastic tubing which also contributed volatile contaminants to the gel when the tubing was used in a suction technique for the removal of silica gel sections from the developed plate for further analysis.

In our experimental studies, five different commercially available silica gels, with and without calcium sulfate binder, were found to be contaminated with organic materials which would confuse the interpretation of the gas chromatographic data. Some of the commercial gels were received in plastic bottles, and some were received in aluminum bottles. The contaminants from the gels packed in the aluminum containers, with plastic caps, were less than the amounts found in the plastic-packed gels, but great enough to cause interpretative problems with the analytical data. Experi-

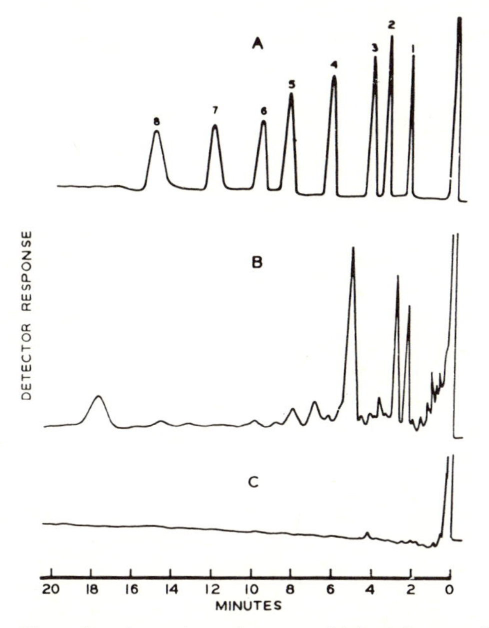

Fig. 1. Gas chromatograph curves obtained from a 1/8 in. × 5 ft. glass column containing 4% SE-30–6% QF-1 silicones on Chromosorb W, HP, 80/100 mesh; column temperature 180°; electron capture detector. (A) Chlorinated pesticide standards: Peaks 1 and 6 are Lindane and Dieldrin, respectively, each 0.3 ng. Peaks 2, 3, 4, 5, 7, 8 are Heptachlor, Aldrin, Heptachlor epoxide, DDE, DDD, DDT, respectively, each 0.6 ng. (B) Contaminants extracted from silica gel prior to heat treatment. (C) Extract of heat-treated silica gel.

ments showed that the plastic containers were at least a partial source of the gel contamination, which confirmed the observations of GEISS *et al.*[4].

Heat treatment of the silica gels at 300° for 16 h effectively removed the contaminants (See Fig. 1); this treatment did not affect the TLC properties of the gels. A less convenient but effective procedure for the removal of contaminants from silica gels is the Soxhlet extraction of the gel with redistilled chloroform for 3 h, followed by extraction with redistilled hexane for 4 h. The Soxhlet cellulose thimbles, if used as extraction containers, must be prewashed in a similar manner.

REFERENCES

1 W. L. LAMAR, D. F. GOERLITZ AND L. M. LAW, *Geological Survey Water-Supply Paper 1817-B*, U.S. Dept. Interior, Washington, D.C., 1965.
2 *FWPCA Method for Chlorinated Hydrocarbon Pesticides in Water and Wastewater*, U.S. Dept. Interior, Washington, D.C., April 1969.
3 R. AMOS, *J. Chromatog.*, 48 (1970) 343.
4 F. GEISS, A. KLOSE AND A. COPET, *Fresenius' Z. Anal. Chem.*, 211 (1965) 37.
5 T. G. LAMONT AND E. CROMARTIE, *J. Chromatog.*, 39 (1969) 325.
6 M. L. SCHAFER, J. T. PEELER, W. S. GARDNER AND J. E. CAMPBELL, *Environ. Sci. Technol.*, 3 (1969) 1261.
7 J. M. MILLER AND J. G. KIRCHNER, *Anal. Chem.*, 24 (1952) 1480.
8 W. L. STANLEY, S. H. VANNIER AND B. GENTILI, *J. Assoc. Offic. Anal. Chem.*, 40 (1957) 282.
9 D. E. BOWYER, W. M. F. LEAT, A. N. HOWARD AND G. A. GRESHAM, *Biochem. J.*, 189 (1963) 24P.
10 T. L. BROWN AND J. BENJAMIN, *Anal. Chem.*, 36 (1964) 446.
11 M. F. KOVACS, JR., *J. Assoc. Offic. Agr. Chem.*, 46 (1963) 884.
12 M. F. KOVACS, JR., *J. Assoc. Offic. Agr. Chem.*, 48 (1965) 1018.
13 D. SMITH AND J. EICHELBERGER, *J. Water Pollution Control Federation*, 37 (1965) 77.

Polarographic Determination of Traces of Nitrilotriacetate in Water Samples

John P. Haberman

FEW PRACTICAL APPLICATIONS of electroanalytical techniques to environmental analyses have been reported. In this study, nitrilotriacetate (NTA) in sewage and river water samples was converted to a polarographically active complex with In(III) in $1M$ NaCl, $0.1M$ acetic acid (HOAc), and $0.1M$ sodium acetate (NaOAc) buffer. Anion exchange concentration and differential polarography were used to increase the sensitivity. A cation exchange column pretreatment was used as a precaution against interfering cations, and isotope dilution analysis with ^{14}C-labeled NTA (NTA–^{14}C) was used to determine the variability of the concentration step.

Colorimetric methods for NTA analysis based on the bleaching of colored complexes have been described (*1*, *2*). Although sensitive and rapid, they are subject to interference by other chelating agents. Colored material in samples may also interfere. Theoretically, other chelating agents could cause polarographic interference (*3*), but in this study using In(III) there was no interference noted.

(1) J. E. Thompson and J. Duthie, *J. Water Pollut. Contr. Fed.*, **40,** Pt. 1, 306 (1968).
(2) R. D. Swisher, M. M. Crutchfield, and D. W. Caldwell, *Environ. Sci. Technol.*, **1,** 820 (1967).
(3) W. Hoyle, I. P. Sanderson, and T. S. West, *J. Electroanal. Chem.*, **2,** 166 (1961).

EXPERIMENTAL PROCEDURES, MATERIALS AND EQUIPMENT

The following steps constituted the procedure used for the analysis of NTA added to waste water samples. Sample collection and storage were not investigated. Figure 1 is a schematic representation of the complete procedure.

Analytical Procedure below 2.57 ppm. INITIAL PRETREATMENT. One liter of filtered sample was measured into a beaker and analytical studies were performed by adding known amounts of NTA to samples at this point. An aliquot of NTA–^{14}C was added for isotope dilution analysis and the pH was adjusted to 3.0 by the dropwise addition of 6*M* HCl. Nitrogen gas was bubbled through the sample with a gas dispersing tube for 10 minutes at a rate sufficient to give good stirring to remove H_2S.

CATION EXCHANGE PRETREATMENT. The sample was passed through a 60-ml (approximately 1.8 × 26 cm) column of cation exchange resin. After the sample had passed through the column, it was rinsed with 200 ml of distilled water, and this rinse was added to the sample. The flow rate of the column was adjusted so that approximately 1 hour was required for passing the sample and rinse through the column. The resin was discarded after each sample.

ANION EXCHANGE ABSORPTION. The pH of the sample was adjusted to 7.0 by the dropwise addition of 50% NaOH solution and the sample was passed through a 15-ml (approximately 1.4 × 12 cm) column of anion exchange resin. The flow rate was adjusted so that approximately 1 hour was required for the 1.2-liter sample to pass through the column. The column was then rinsed with 100 ml of distilled water, and all column effluent was discarded.

ELUTION OF ANION EXCHANGE COLUMN. The NTA was stripped from the column with 1*M* NaCl in 0.1*M* NaOAc and 0.1*M* HOAc buffer of pH 4.7. The flow rate was adjusted to approximately 10 ml per minute and the first 10-ml fraction contained most of the distilled water from the free volume of the resin, so it was discarded. The next 10-ml fraction contained approximately half of the NTA and the next 50-ml fraction contained almost all of the rest of the NTA. The largest concentration factor was obtained by using the second 10-ml fraction for analysis. The anion exchange resin was discarded after each sample.

ISOTOPE DILUTION ANALYSIS. A small portion of the fraction of eluant (100 μl) was pipetted directly into a scintillation vial for counting. The recovery of NTA–^{14}C compared to the amount added in the "Initial Pretreatment" step was used to calculate the concentration factor.

POLAROGRAPHIC ANALYSIS. The rest of the eluant fraction was used for differential polarographic analysis (the eluant served as the supporting electrolyte solution). Three milliliters were pipetted into the sample and reference cells. Both sample and reference solutions were deaerated by bubbling nitrogen gas through the sample in the cells for 5 minutes. A potential of −0.70 volt *vs.* SCE was applied with the current sensitivity at 0.05 μA full scale. $In(NO_3)_3$ (0.001*M*) in 1*M* NaCl, 0.1*M* NaOAc, and 0.1*M* HOAc (pH 4.7) was added to the sample side of the cell with a 1-ml microburet. After each addition, nitrogen gas was bubbled through the solution for stirring [and removal of dissolved oxygen introduced *via* the In(III) solution] and then stopped. When a polarographic current of 0.005–0.02 μA indicated the presence of excess In(III), addition was stopped and the volume of In(III) solution added to the cell was noted. Note that for some sewage samples, it was necessary to finish "titrating" the sample with 0.01*M* $In(NO_3)_3$. A preliminary potential sweep was manually performed to determine the current sensitivity setting which resulted in the best recorder display. Then a cathodic polarographic sweep was performed from an initial potential of −0.10 volt *vs.* SCE and a polarogram was recorded.

The height of the excess In(III) wave was used as the base line to determine the height of the In(III)–NTA wave and the current was measured at the peak of the wave (Figure 2). It was necessary to correct the polarographic result for dilution by the volume of In(III) solution added. The current was correlated to the solution concentration of NTA as ppm Na_3NTA *via* a standard curve.

Analytical Procedure above 2.57 ppm. The "Initial Pretreatment" step was the same except that solutions of 100-ml volume were used and NTA-^{14}C was not added. All other steps until the "Polarographic Analysis" were deleted.

Materials and Equipment. Four types of waste water sample were used; Ohio River water was collected at Cincinnati and used without further treatment. Cincinnati municipal sewage (Gest Street Plant) was also used without further treatment. Synthetic sewage prepared from a glucose nutrient broth and dipotassium phosphate was treated under laboratory conditions by activated sludge (aerobic). Synthetic sewage with peptone, glucose, urea, disodium phosphate, sodium chloride, and beef extract was treated under laboratory conditions in sludge digesters (anaerobic).

Samples were suction filtered through two thicknesses of glass filter paper. The unlabeled NTA was Matheson Coleman and Bell reagent grade nitrilotriacetic acid and the NTA–^{14}C was an Amersham/Searle Corp. custom preparation with one carboxyl ^{14}C per molecule. Dowex 50 W-X8 (50–100 mesh) cation exchange resin in the hydrogen ion form was used throughout. Dowex 1-X8 (50–100 mesh) anion exchange resin was used for early experiments and Dowex 21 K (50–100 mesh) was used for later experiments including evaluation with waste water samples (both in the chloride ion form). Both cation exchange and anion exchange resins were prewashed with two cycles involving an HCl wash and distilled water rinse, then were rinsed to approximately neutral pH with water. In(III) solutions were prepared by dissolving a weighed quantity of American Smelting and Refining Co.

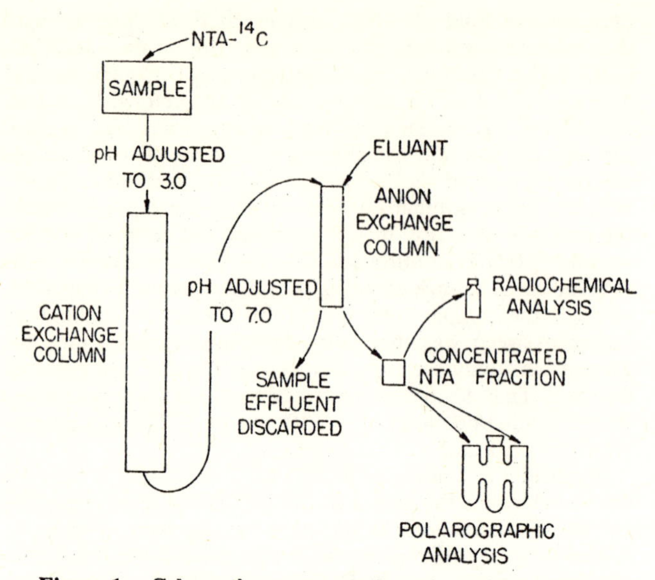

Figure 1. Schematic representation of complete analysis

99.999% indium metal with dropwise additions of concentrated HNO_3 while heating on a steam bath. All other chemicals were reagent grade, and used without further purification. Radioanalytical determinations of NTA–^{14}C were performed with a Packard Tri-Carb Model 3375 liquid scintillation spectrometer. The differential Polarograph (*4*) was a simple one and was programmed on a versatile electrochemical instrument. Details are not given here since a wide variety of commercial electrochemical instruments are available (*5*).

DISCUSSION

Sample collection and storage were not investigated, but if the NTA–^{14}C were added to unfiltered sample at the time of collection, loss of unknown NTA during sample storage would be automatically corrected for if it was in equilibrium with the NTA-^{14}C added.

Ion Exchange of NTA Samples. Cation Exchange Removal of Cations. The pH of samples was adjusted to 3.0 in the initial pretreatment step so that the ionic strength introduced was not appreciable while the pH was low enough to

(4) L. Meites, "Polarographic Techniques," 2nd ed., Interscience Publishers, New York, N. Y., 1965, p 573.
(5) G. W. Ewing, *J. Chem. Educ.*, **46**, A717 (1969).

facilitate the displacement of metal ions from NTA complexes (high ionic strength could interfere with the absorption of NTA during the later anion exchange step). The hydrogen ions liberated from the column when the metal ions were absorbed lowered the pH further, with no increase in the ionic strength (after cation exchange treatment, the pH of most samples was in the range of 1.5 to 2.5). The effectiveness of the cation exchange step in removing metal ions was not determined with actual samples but Zn(II) and Mg(II) in the presence of NTA were reduced by a factor of 10^5 in distilled water.

ANION EXCHANGE ABSORPTION OF NTA. Figure 3 illustrates the absorption of NTA–^{14}C onto Dowex 1-X8 resin. NTA is a divalent anion in the pH range of 5–8 (*6*) and a pH of 7.0 was selected for most experiments, but absorption was efficient in the pH range of 5–11. The NTA was taken up by the first part of the resin in spite of the fact that 1 liter of sample was passed through the column in each case.

The optimum size of anion exchange column was not determined during this work. Smaller volumes of resin in the anion exchange column required less eluant to elute the NTA and resulted in higher concentrations of NTA (note that this consideration did not apply to the cation exchange pretreatment step).

ELUTION OF NTA. Figure 4 shows the elution behavior of NTA with sewage samples. Some variability in the elution of NTA from different types of samples was evident from these studies. One of the reasons for selecting anion exchange as a method of concentrating NTA was to have the possibility of a chromatographic elution to separate interferences (*7*). No evidence of interference during the polarographic analysis step was found in this study so the practice of using a high concentration of salt to strip the NTA (and presumably other materials) from the column was used. It was found that concentrations of NaCl of 1*M* (and greater) eluted the NTA as soon as the free volume of the resin was displaced. NTA was efficiently stripped from Dowex 21 K resin in this way even when the resin was visibly darkened from material in sewage samples.

Table I is a summary of recoveries observed during an evaluation of the analysis. The first 10-ml fraction contained very little NTA–^{14}C. An average concentration factor of 47 resulted from the second 10-ml fraction and an average concen-

---

(6) "Stability Constants of Metal-Ion Complexes," L. G. Sillen and A. E. Martell, Ed., Chem. Soc. (London), Spec. Publ. No. 17, 1964.

(7) C. Davies, R. D. Hartley, and G. J. Lawson, *J. Chromatogr.*, **18**, 47 (1965).

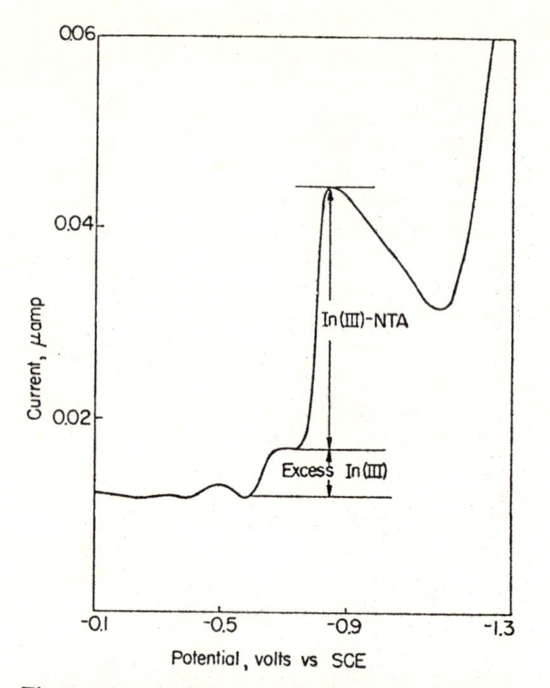

Figure 2. Activated sludge-treated synthetic sewage with 0.0257 ppm of NTA (plus 0.0257 ppm from NTA-^{14}C to give a total of 0.0514 ppm)

It was necessary to add approximately a 30-fold molar excess of In(III) to total NTA before an excess In(III) wave appeared

tration factor of 15.5 resulted from a total volume of 60 ml of eluant (not including the first 10-ml fraction). Note that the relative standard deviation (RSD) of the concentration factor in the 60-ml fraction was only 7% while that of the 10-ml fraction was 12%. If it was established in practice that the reproducibility of recovery of a particular fraction was well within the experimental error of the whole procedure, then it would be possible to discontinue the isotope dilution step.

Polarographic Analysis of NTA as In(III)–NTA. Other metal cations were investigated for the polarographic analysis of NTA but In(III) gave the best results during preliminary studies on sewage samples. Without regard for the actual species involved in the equilibrium system and the reduction reactions, they may be represented as follows:

$$\text{In(III)} + \text{NTA} \rightleftharpoons \text{In(III)}-\text{NTA}$$

$$\text{Excess In(III)} + 3e^- \rightarrow \text{In(0)} \qquad -0.61 \text{ volt } vs. \text{ SCE}$$

$$\text{In(III)–NTA} + 3e^- \rightarrow \text{In(0)} + \text{NTA} \qquad -0.79 \text{ volt } vs. \text{ SCE}$$

The polarographic wave from excess In(III) in $0.1M$ acetic acid and $0.1M$ sodium acetate (pH 4.7) exhibited a half-wave potential of -0.61 volt *vs.* SCE. The In(III)–NTA reduction wave was well separated from the excess In(III)

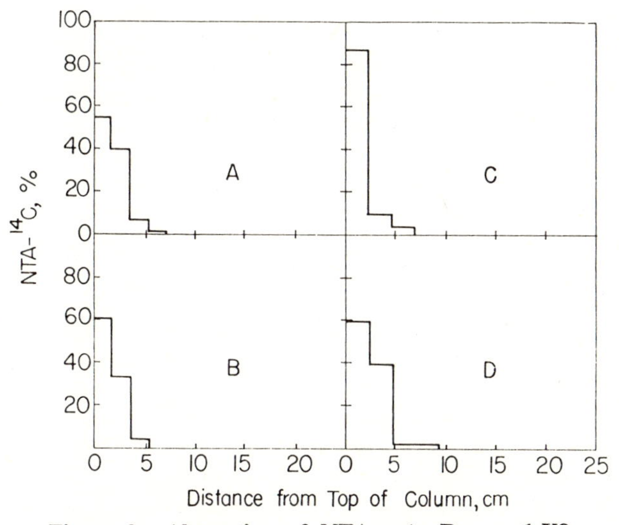

Figure 3. Absorption of NTA onto Dowex 1-X8

A. **25.7 ppm of Na_3NTA in distilled water at pH 11.0; 0.1% of sample in column effluent**

B. **25.7 ppm Na_3NTA with 10-fold molar excess of Pb(II) in distilled water at pH 5.0; 0.9% of sample in column effluent; absorption was as a Pb(II) complex**

C. **25.7 ppm Na_3NTA in tap water with 7 grain hardness at pH 7.0; pretreated with Dowex 50 W-X8; 0.7% of sample in column effluent**

D. **0.257 ppm NTA under same conditions as *C.* above; $\leq$ 3% of sample in column effluent (limit of detection of available NTA-^{14}C)**

Table I. Average Recovery of NTA-^{14}C during Evaluation

Fraction, No., ml	Recovery, %	RSD, %
2, 10	46	12
3, 50	47	15
2 + 3, 60	93	7

wave and exhibited an unusual peak shape (*8*). The half-wave potential was found to be −0.79 volt *vs.* SCE and the peak potential was −0.85 volt in the acetate buffer. Apparently the reduction was quasi-reversible and the electroactive species was In(III) $HNTA^+$ (*9*). Polarographic studies of other In(III) complexes which exhibited peak shaped waves have been reported in the literature (*10, 11*). In practice, the peak shape of the waves increased the selectivity of the determination because the peak represented an excellent reference point at which to make the current measurement. Standard curves based on the peak current were linear in appearance but the ratio of the peak current to the NTA concentration changed slightly with the NTA concentration.

The conversion of the NTA to the In(III) complex was an extension of a technique which has already been described (*3*). It was controlled by controlling the amount of excess In(III) in equilibrium with the complex. This was done by adding In(III) to the sample cell until the current at −0.70 volt *vs.* SCE indicated the presence of approximately the same amount of excess In(III) for both calibration standards and unknown samples. Figure 5 illustrates the relatively small effect of the amount of excess In(III) on the reduction current for In(III)–NTA. The horizontal axis represents the excess In(III). The vertical axis represents the magnitude of the reduction current of the In(III)–NTA complex compared to the current when the magnitude of the In(III) wave was in the range used as standard conditions for analyses (0.005–0.02 μA). The points representing standard conditions for each concentration are the symbols centered on the 0% error line. The height of the In(III)–NTA wave was within ±10% of standard conditions for a large range of excess In(III) (0.4 to 100% with 257 ppm Na_3NTA). This was probably due to competing equilibrium and kinetic (*12*) effects.

The relative standard deviation of standards was approximately 5% (with and without temperature control of the cell). The lower limit of NTA which could be measured above background currents (including the current from the excess In(III) wave) was about 0.3 ppm when anion exchange concentration was not used and the measurement was made with a differential polarograph.

---

(8) J. Koryta and I. Kössler, *Collection Czech. Chem. Commun.*, **15**, 241 (1950).
(9) R. Staroscik and K. Webs, *Chem. Anal.* (*Warsaw*), **12**, 1275 (1967).
(10) A. J. Engel *et al.*, ANAL. CHEM., **37**, 203 (1965).
(11) L. Pospisil and R. DeLevie, *J. Electroanal. Chem.*, **25**, 245 (1970).
(12) C. Auerbach, ANAL. CHEM., **30**, 1723 (1958).

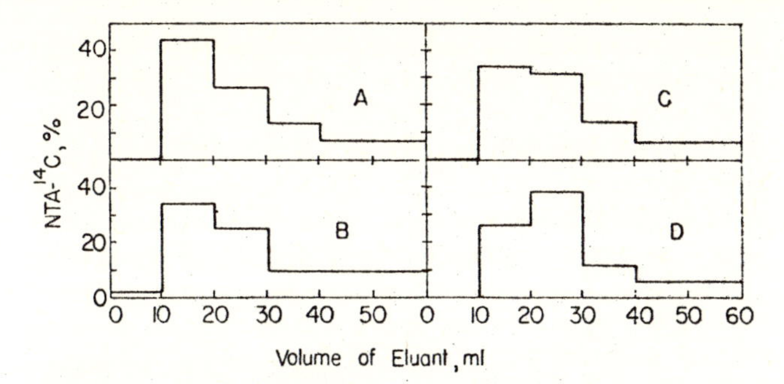

Figure 4. Elution of NTA from Dowex 21 K with 1M NaCl, 0.1M HOAc, and 0.1M NaOAc (pH 4.7)

_A_. 0.0257 ppm of Na_3NTA in distilled water; 89% recovery in 60 ml of eluant

_B_. 0.283 ppm of Na_3NTA in untreated municipal sewage; 79% recovery in 60 ml of eluant

_C_. 0.283 ppm Na_3NTA in activated sludge treated synthetic sewage; 86% recovery in 60 ml of eluant

_D_. 0.283 ppm Na_3NTA in sludge digester treated synthetic sewage; 87% recovery in 60 ml of eluant

Interference Studies. No evidence of interference was detected in distilled water solutions with a tenfold molar ratio of ethylenediaminetetraacetate (EDTA), ethane-1-hydroxy-1, 1-diphosphonate (EHDP), citrate, maleate, phosphate, carbonate or sulfate to 2.57 ppm Na_3NTA. A solution containing EHDP and EDTA, each with a tenfold molar ratio to NTA resulted in no interference as was also the case for a solution containing all of these compounds, each with a tenfold molar ratio to NTA. Although it was often necessary to add much more In(III) than was necessary for the NTA alone, before excess In(III) was detected, the polarograms were not affected and the analytical results were within experimental error of being the same in these studies.

Anomalous results were observed when In(III) was added to a distilled water sample which had been treated with a particular batch of Dowex 21 K resin. The resin had been left in contact with concentrated HCl for 5 hours during pretreatment. When the resin was retreated with 1M HCl in contact with the resin for a short time, normal results were obtained. Benzyltrimethylammonium chloride was added to NTA in distilled water to represent a possible decomposition product of Dowex 21 K resin, but the anomalous results were not reproduced. The wave shapes for the free In(III) and In(III)–NTA were normal but the height of the In(III)–NTA wave was decreased 29% when 200 ppm of benzyltrimethylammonium chloride was present and 41% when 400 ppm was present. In

Table II. Evaluation with Anion Exchange Concentration

Sample	NTA added, ppm Na_3NTA	Analysis of Fractions			NTA found, ppm Na_3NTA	Error, %
		Fraction, No., ml	NTA–^{14}C, %	Total NTA, ppm Na_3NTA		
River water	0.0257	2, 10	47	2.38	0.0249	−3
	0.0257	3, 50	61	0.395	0.0072	−72
	0.257	2, 10	56	14.5	0.233	−9
	0.257	3, 50	41	1.98	0.216	−16
	2.57	2, 10[a]	44	11.2	2.52	−2
	2.57	3, 50	55	26.2	2.35	−9
Activated sludge effluent	0.0257	2, 10	46	2.02	0.0182	−29
	0.0257	3, 50	45	0.452	0.0245	−5
	0.257	2, 10	51	12.3	0.215	−16
	0.257	3, 50	45	2.35	0.235	−9
	2.57	2, 10[a]	52	12.6	2.39	−7
	2.57	3, 50[a]	42	2.10	2.47	−4
Raw sewage	0.0257	2, 10	36	1.45	0.0146	−43
	0.0257	3, 50	52	0.585	0.0305	+19
	0.257	2, 10[a]	38	1.05	0.250	−3
	0.257	3, 50	51	2.48	0.217	−16
	2.57	2, 10[a]	44	11.8	2.68	+4
	2.57	3, 50[a]	50	2.53	2.50	−3

[a] Fraction diluted 10-fold before polarographic analysis.

practice, materials of this type which might occur in samples should be efficiently removed during cation exchange treatment.

The polarographic reduction of In(III)–NTA did not appear to be particularly sensitive to interference by surfactants. Linear alkyl benzene sulfonate concentrations of 3.5 and 43.5 ppm did not affect polarograms and an analysis for NTA in a sample of a commercial detergent formulation dissolved in electrolyte solution gave a value which was within experimental error of the expected result.

EVALUATION. The experiments summarized in Tables II and III were performed to evaluate the best set of conditions which evolved from this work. One-liter portions of a river water sample, a laboratory activated sludge treated sewage sample, a raw municipal sewage sample, and a laboratory sludge digester-treated sample were carried through the complete procedure with 0.0257, 0.257, and 2.57 ppm of unlabeled NTA added. The second 10-ml fraction of the anion exchange eluant and a third 50-ml fraction were analyzed. The accuracy of the analytical results for the two fractions was approximately the same. The relative standard deviation of averaged results was 42, 6, and 5%, respectively, for 0.0257, 0.257, and 2.57 ppm Na_3NTA. The analytical results for the digester sample were high but when the NTA added to the samples was subtracted out, a relatively constant blank of 4.0 ppm (13% RSD) was observed. Although contamination was not expected, it was considered more likely that this sample had become contaminated with NTA than that an interference had been encountered. Analytical data from the di-

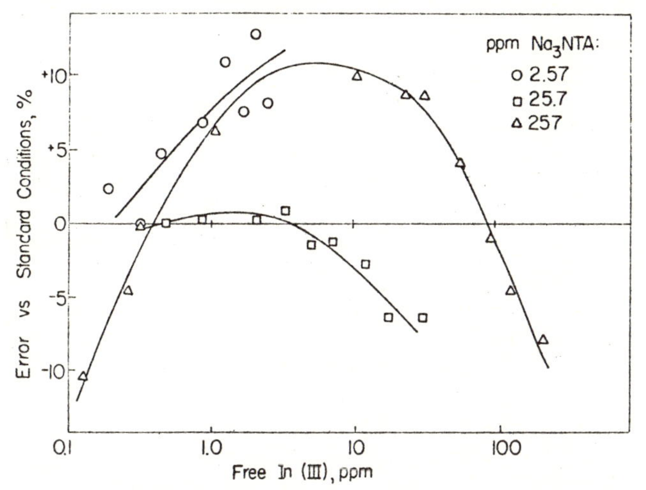

Figure 5. Effect of excess In(III) on height of In(III)–NTA wave

Table III. Evaluation without Anion Exchange Concentration

Sample	NTA added, ppm Na_3NTA	NTA found, ppm Na_3NTA	Error, %
River water	2.57	2.49	−3
	25.7	25.9	+1
	257	239	−7
Activated sludge effluent	2.57	2.55	−1
	25.7	25.3	−2
	257	229	−11
Raw sewage	2.57[a]	2.29	−11
	25.7[b]	25.2	−2
	257[b]	252	−2
Digester effluent	2.57	2.64	+3
	25.7[b]	31.2	+21
	257[b]	267	+4

Sample diluted 5-fold[a] or 10-fold[b] before polarographic analysis.

gester effluent sample were not included in the results on Table II. The results of a second series of the same type of samples (collected at a different time) with 2.57, 25.7, and 257 ppm NTA added are presented in Table III. No NTA–^{14}C was added and all ion exchange steps were deleted. The relative standard deviation of averaged results was 6, 11, and 7%, respectively, for 2.57, 25.7, and 257 ppm Na_3NTA. The results indicate that In(III) has the ability to displace NTA from other metal ions naturally present in these types of samples without the necessity of cation exchange pretreatment. In this series of experiments, the digester sample did not give anomalously high results and analytical data from the digester sample were included with the other types in Table III. In both series of experiments, a precipitate sometimes appeared upon addition of In(III) solution to the sample in the polarographic cell. In order to obtain normal polarographic results, it was then necessary to dilute the sample with electrolyte solution sufficiently that precipitation did not occur.

The polarographic residual current was higher than normal at negative potentials with sewage samples and this effect decreased the negative potential range for reductions at a DME but did not interfere with the reduction of the In(III)-NTA complex (see Figure 2). The analytical results at the 0.0257-ppm level were marginal but the results at the 0.257-ppm level and greater were much better. The NTA–^{14}C added for isotope dilution data was 0.0257 ppm (1 mCi/g specific activity) for all of the results with anion exchange concentration. At the 0.0257-ppm level of unlabeled NTA, the presence of the NTA–^{14}C had the effect of doubling the error in determining the concentration of the unlabeled NTA (which would correspond to the unknown amount in an actual analysis).

The use of a smaller amount of a "hotter" sample of NTA-^{14}C would be expected to improve analysis at that level considerably.

Before the excess In(III) wave appeared, it was often necessary to add considerably more In(III) than was necessary to complex the NTA in the representative samples (see Figure 2 caption). Apparently there were large quantities of material in these samples which formed complexes with In(III) which were not polarographically active in the electrolyte used. As long as excess In(III) was present, the presence of polarographically inactive In(III) complexes did not seem to affect the reduction of the In(III)-NTA complex.

For these studies, differential polarography was used as a more sensitive method than dc polarography. Differential polarography also had the advantage of subtracting out residual currents which might be encountered. In practice, there was no evidence of interfering residual currents in samples and there was no evidence that single-cell polarographic techniques could not be used. There was evidence that organic material in sewage samples interfered with the use of stationary mercury drop electrodes and that techniques employing dropping mercury electrodes were to be preferred.

There was no evidence that NTA was lost during the cation exchange step for these studies but large losses were encountered during further development of the method for septic tank samples and the cation exchange step was deleted (*13*). The suggestion was made that better recovery might be obtained with the cation exchange resin in the sodium ion form. The "Initial Pretreatment" step was also discontinued for septic tank samples.

ACKNOWLEDGMENT

The assistance of D. F. Kuemmel and T. R. Williams is gratefully acknowledged.

---

(13) J. E. Thompson, Sanitary Engineering Research Laboratory, Procter & Gamble Co., Cincinnati, Ohio, personal communication, 1970.

ANALYSIS OF WATER FOR MOLECULAR HYDROGEN CYANIDE

K. H. Nelson and I. Lysyj

Knowledge of the relationship of hydrogen cyanide (HCN) in the toxicity of water-borne cyanide salts to aquatic life is important in toxicological evaluation of waste disposal in waterways. In acute toxicity studies of alkali cyanides to fish, Wuhrman and Woker [1] found molecular HCN to be the toxic factor. For various complex metal cyanides in water, Doudoroff *et al.*[2] established the direct relationship between acute toxicity to fish and the concentration of undissociated HCN. They found the toxicity of metal cyanides to be independent of total cyanide levels but related to the HCN concentration in the system. Other studies have indicated that the temperature and oxygen content of the water affect the susceptibility of fish to HCN. In support of studies of this nature, specific and sensitive analytical procedures are necessary for the determination of undissociated HCN to the exclusion of other cyanide forms in the water.

Gas chromatographic methods based on the analysis of the equilibrated vapor phase in contact with the aqueous solution containing all the cyanide compounds have been reported. Both thermal conductivity [3] and flame ionization [4] detectors were used. In principle, the methods are specific for undissociated HCN in solution provided the volume of vapor taken for analysis relative to the solution volume is not so large as to deplete appreciably the molecular HCN concentration in the solution and thus shift the various chemical equilibria. However, the use of a desiccant for removal of the water vapor from the gaseous sample and the presence of natural volatile organics in some waters may present problems in the final step of HCN measurement.

Microgram quantities of HCN have been determined by measuring the spontaneous electrolysis current flowing between a silver anode and a platinum cathode.[5] Also, small amounts of cyanide have been measured potentiometrically with silver and gold anodes.[6] Based on these studies, McCloskey [7] demonstrated direct amperometry of cyanide at extreme dilutions in a system with a small potential applied between a rotating silver anode and a stationary platinum cathode. Using a rotating gold anode, Miller *et al.*[8] recently extended the technique into the submicromicrogram range of cyanide concentration.

When combined with vapor phase equilibration, the sensitive amperometric technique offers a highly specific and sensitive method for the measurement of small amounts of molecular HCN in water. The method presented here consists of sparging a small portion of the undissociated HCN from the water sample, trapping the HCN in dilute base, and then measuring the sparged HCN with a rotating gold anode.

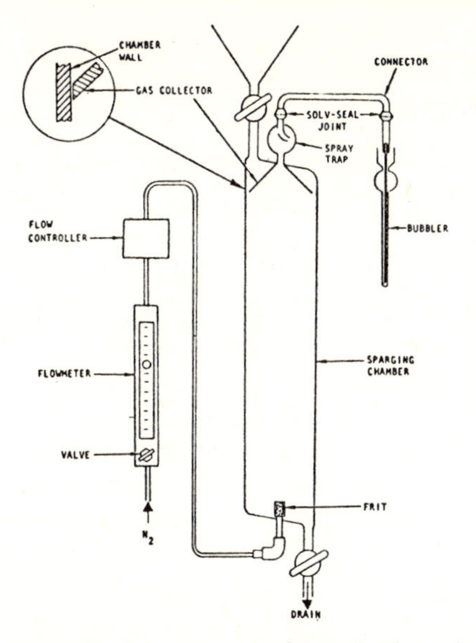

FIGURE 1.—Apparatus for collection of molecular HCN from water.

Apparatus

The apparatus used to collect the molecular HCN from the water sample is shown in Figure 1. It consisted of four components mounted on a rack. These were the sparging chamber, a bubbler, a flow controller, and a flowmeter with a shut-off valve.

Sparging Chamber

The glass sparging chamber was 70-mm outside diameter (OD) by 68-cm high with a capacity of 2 l of water. At the bottom, a 6-mm bore stopcock with a Teflon plug was used to drain the chamber. A coarse-porosity frit, which dispersed the nitrogen, was sealed to the chamber bottom and terminated externally in a 6.35-mm OD glass tube. This tube was connected, by means of a reducer * and an elbow union† fitted with a polyethylene front ferrule, to the 0.125-in. (0.318-cm) OD metal line from the flow controller ‡ and the flowmeter. At the top, a funnel-shaped gas collector was sealed to the spray trap which was fastened to a 5-mm joint.§ The clearance between the gas collector and the chamber walls, shown in the insert in Figure 1, allowed the water sample to fill the chamber without any reduction in the efficiency of collection of the gas bubbles during sparging. The water sample was introduced into the chamber by means of the 5-in. (12.7-cm) funnel sealed to the 6-mm oblique bore stopcock on top of the chamber.

Bubbler

The bubbler was composed of two parts. The outer part, for containment of electrolyte, was a 15-cm length of 5-mm diam glass tubing fastened to a 2-cm diam spherical section which was joined to a 3-cm length of 17-mm diam tubing. The inner part was a stainless steel tube, 8.5 in. (21.6 cm) long with an OD of 0.0625 in. (0.159 cm) having a bore of 0.023 in. (0.06 cm), epoxyed in a 5-mm joint.§ The bubbler was connected to the spray trap by means of a glass tube fabricated from two joints.§ The glass tube was warmed with either a heating tape or a heat gun.

Polarograph

Measurement of the sparged HCN was made with a polarograph.‖ The glass cell, 6.5 mm by 17 mm inside diameter (ID), had a stationary platinum cathode and a gold anode that was rotated at a constant speed of 600 rpm by a hollow-spindle synchronous rotator.#

Cathode

The platinum cathode was fabricated by spot welding a platinum wire lead

* Swagelok 200-R-4.
† Swagelok 400-9.

‡ Brooks Instruments, Model 8743 Elf.
§ Solv-Seal glass/Teflon.
‖ Sargent Model XXI. Any polarograph with similar capabilities can be used.
Sargent, S-76485.

to a 2-cm by 2-cm platinum foil which was then contoured to fit the cell wall.

Anode

The gold anode ** was tripled in area by cutting the existing wire near the upper terminus and spot welding additional gold wire between the cut ends. The entire wire was helically wound within the original length on the glass electrode tube. The electrical connection to the gold anode, shown in Figure 2, was constructed by first soldering a nickel wire to the cap of the electrode. After the electrode was positioned in the synchronous rotator, the glass tube was attached to the electrode with the rubber tubing. Then a small amount of epoxy cement was placed over the solder by means of a syringe. The mercury was added after the cement hardened.

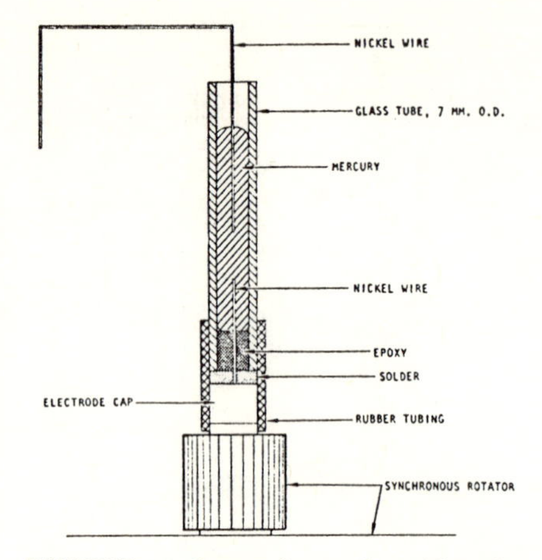

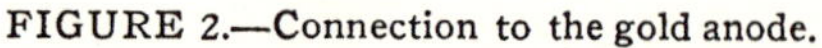

FIGURE 2.—Connection to the gold anode.

Electrode Cleaning

The electrodes were cleaned by immersing them in 1:1 nitric acid, washing them with distilled water, and then immersing them in 0.03-*M* NaOH until used. Between analyses and for overnight storage, the cell and electrodes were first rinsed with distilled water. Then the electrodes were immersed in 0.03-*M* NaOH contained in the cell.

The cost of the equipment as modified without the polarograph would be about $300. A suitable polarograph would cost in the range of $1,500 to $3,500.

Procedure

Cleaning

Before assembly, clean all glass apparatus parts with a dichromate-sulfuric acid solution. Rinse thoroughly with distilled water and then assemble. Before analyzing each sample, rinse the sparging chamber by first filling with distilled water to the top of the spray trap. Then close the spray trap outlet and pass nitrogen into the chamber to displace the air above the gas collector with water. This air escapes through the water in the filling funnel. Discontinue the nitrogen flow and open the spray trap outlet to allow water to refill the spray trap. Then drain the chamber. Dry the spray trap by blowing hot air on the exterior. Rinse the connector and both parts of the bubbler with distilled water and dry.

Sparging

To sparge HCN from a sample, first introduce 2 l of sample into the sparging chamber. The liquid level should be about 0.25 in. (0.64 cm) above the bottom edge of the gas collector. Using a 1-cc tuberculin syringe with a 6-in. (15.2-cm) long 18-gauge needle, place 1 cc of 0.07-*M* NaOH in the outer part of the bubbler. Assemble the bubbler and connect it to the spray trap. Then pass 1 l of nitrogen through the sparging chamber at a rate of 35 cc/min.

Cell Preparation

During the sparging, prepare the cell for measuring the sparged HCN. Rinse the cell and electrodes with distilled water and then pipette 10 cc of 0.03-*M* NaOH into the cell. Start rotation of the gold anode and apply a potential of 300 mv across the electrodes.

** Sargent, S-30435.

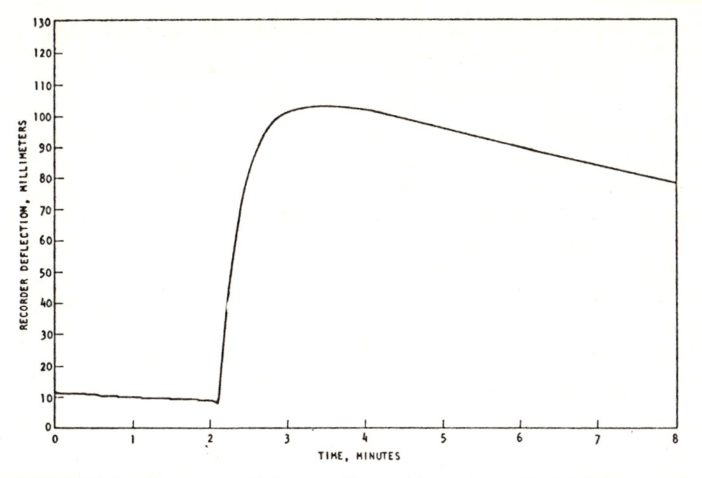

FIGURE 3.—Response of the rotating gold anode to the HCN sparged from a 9-ppb solution.

The polarograph controls are set with damping on 1, applied EMF at 30 percent of a 1-v span, and sensitivity of 0.003 μa/mm or as required. The initially large residual current will decay within a few minutes to lower values and finally attain a negligible decay rate.

Measurement of Collected Hydrogen Cyanide

After the sparging is completed, stop the nitrogen flow and remove the outer part of the bubbler. Then transfer the NaOH solution from the bubbler to the cell, which now has a negligible residual current. The transfer is accomplished with a 1-cc tuberculin syringe equipped with an 8-in. (20.3-cm) long 18-gauge needle. Record the volume of solution transferred. The current will rise quickly, remain at a peak value for several seconds, and then slowly decay (Figure 3). It is important that the sample be rapidly introduced from the syringe and be thoroughly mixed with the cell electrolyte before the recorder pen reaches its peak response. On the chart, measure the height between the peak response and the residual current value before addition of the bubbler solution.

Calculation

To calculate the concentration of HCN in the water, multiply the measured response height on the polarographic chart by the instrument sensitivity and the fraction of NaOH solution transferred from the bubbler to the cell. After correcting this current for an experimentally determined blank, read the nanograms of HCN from the calibration chart. This value together with the distribution constant for HCN permits calculation of the parts per billion HCN present in the water sample. The distribution constant [6] for HCN between the vapor phase and the water is 3×10^{-3}.

Calibration

Prepare a series of calibration solutions by adding 0, 0.5, 1.0, 1.5, 2.0, and 2.5 ml of 0.0005-*M* potassium cyanide (KCN) solution to 0.07-*M* NaOH in 200-ml volumetric flasks. With fresh 0.03-*M* NaOH in the cell, and after the residual current is a negligible value, add 1 ml of a calibration solution to the cell. Measure the peak height from the initial baseline to the maximum response. Then multiply the peak height by the instrument sensitivity setting to obtain the current in μa. This procedure is repeated for each solution to establish the calibration chart (Figure 4) for the cell response of μa vs. nanograms of HCN. As a daily calibration check, use a freshly prepared solution containing the 2 ml of the 0.0005-M KCN.

Results

In this investigation, analytical reagent grade chemicals and boiled distilled water were used throughout. For preparation of the calibration chart and the synthetic samples, a 0.0005-M KCN solution was prepared daily by diluting 5 ml of a 0.1-M stock solution, prepared weekly, to 1 l with distilled water. To adjust the pH of the synthetic samples, increments of two stock buffer solutions were used. These solutions were a 0.15-M KH_2PO_4 solution, prepared by diluting 20.41 g of the salt to 1 l, and a 0.15-M Na_2-

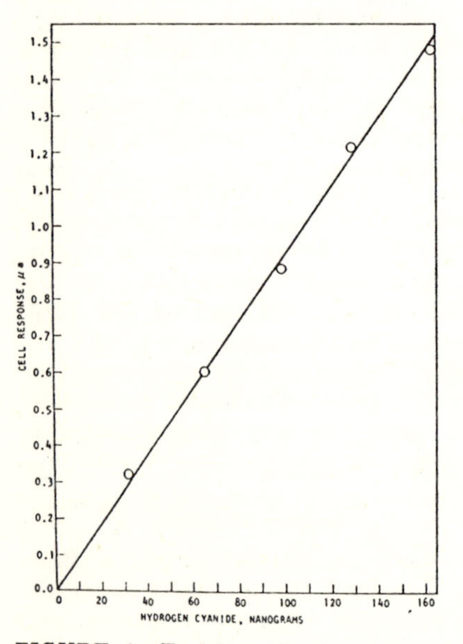

FIGURE 4.—Typical calibration chart of cell response vs. nanograms of HCN.

TABLE I.—Analysis Results for Synthetic Unknowns Containing Added HCN

pH 6.1		pH 7.2		pH 8.2	
Present (ppb)	Found (ppb)	Present (ppb)	Found (ppb)	Present (ppb)	Found (ppb)
2	2	5	7	5	3
2	4	5	8	5	3
5	5	20	23	9	10
5	6	20	21	9	10
10	9	20	16	18	23
10	10	20	19	18	21
20	23	50	50	46	46
20	20			46	41
20	23			46	42

HPO_4 solution containing 21.29 g of anhydrous Na_2HPO_4 in 1 l of distilled water.

In developing the method, synthetic unknowns containing various amounts of HCN were prepared with the distilled water. An unknown was perpared by introducing the required volumes of stock buffer solutions from burettes into a 2-l volumetric flask containing about 1 l of boiled, distilled water, adding the desired volume of 0.0005-M KCN solution from a burette, and then diluting to the mark with the distilled water. The pH of an unknown was either 6.1, 7.2, or 8.2. The amounts of stock buffer solutions used to buffer to these acidities were 90.0, 38.9, and 6.0 ml, respectively, of the KH_2PO_4 solution and 10.0, 61.1, and 94.0 ml, respectively, of the Na_2HPO_4 solution. Each unknown was sparged immediately after preparation, and the results are shown in Table I. The standard deviation is 1.42 for the pH 6.1 solutions, 2.59 for the solutions with a pH of 7.2, and 3.12 for the pH 8.2 solutions. For all the solutions as a group, the standard deviation is 2.36.

After developing the method with the synthetic unknowns, the procedure was applied to natural waters to determine recovery of added HCN. One of these waters was fresh water obtained from Sespe Creek in Ventura County, Calif. This stream, which has

TABLE II.—Analysis of Samples Prepared with Sespe Creek Water (Calif.)

Molecular HCN Present, (ppb)	Molecular HCN Found, (ppb)
4	4
4	5
8	8
8	11
9	9
10	11
10	10
16	14
18	19
45	41
45	44
45	41
46	44
91	82

many small tributaries, drains a large, rugged, uninhabited mountainous area northwest of Los Angeles. The water was transported in 5-gal (18.9-l) glass carboys that were rinsed with the water before filling. After the water had stood a few hours, a small amount of algae settled out. The pH of the water was 8.2. For testing the analytical procedure, known amounts of KCN were added to 2-l portions of the water to give solutions containing the concentrations of molecular HCN listed in Table II. Preparation of these samples consisted of adding appropriate aliquots of 0.0005-*M* KCN solution from a burette to approximately 1 l of the water in a 2-l volumetric flask and then filling to the mark with water. The results for these samples, which were analyzed immediately after preparation, are shown in Table II. The standard deviation for this series of samples is 1.35.

Any substance that has a significant vapor pressure above its solution will be sparged into the bubbler and be present during the measurement of the HCN. In the determination of cyanide ion, Miller *et al.*[8] and McCloskey[7] measured the current response of a number of possible interfering compounds, and of the tested anions, only sulfide would be expected to interfere in the present method. An experimental distribution constant for H_2S is not readily available, but ideal behavior can be assumed in calculating a constant. For water containing 10 ppb H_2S, the mole fraction of H_2S in water is 6×10^{-9}. The vapor pressure of H_2S at 25°C is 20 atm, and, therefore, from Henry's Law the vapor pressure at 10 ppb is 1.2×10^{-7} atm. This is equivalent to 170 nanograms/l. Therefore, a distribution constant can be calculated:

$$K = \frac{\text{mg } H_2S\text{/l of gas}}{\text{mg } H_2S\text{/l of solution}} = \frac{170 \times 10^{-6}}{1 \times 10^{-2}} = 1.7 \times 10^{-2}$$

A comparison of this distribution constant with that for HCN, 3×10^{-3}, shows that approximately six times more H_2S would be sparged from water containing equivalent concentrations of the two substances. Because the anodic response of cyanide is about 100 times that of sulfide,[8] H_2S would not seriously interfere when present at concentration levels equal to or less than the HCN but would begin to introduce an increasingly significant error as the H_2S concentration exceeded that of HCN.

The present detection limits for HCN by this method do not seem to be limited by the sensitivity of the final measurement but rather by the techniques of transferring and handling nanogram quantities of the desired species. With refinements in these techniques, it should be possible to apply the procedure to samples with lesser concentrations of HCN.

Acknowledgment

The research on which this publication is based was performed pursuant to Contract No. 14-12-491, with the FWPCA, U. S. Department of the Interior.

References

1. Wuhrmann, K., and Woker, H., "Experimental Investigations on Ammonia and Hydrocyanic Acid Poisoning." *Schweiz. Zeits. Hydrol.* (Switz.), **11**, 210 (1948).
2. Doudoroff, P., *et al.*, "Acute Toxicity to Fish of Solutions Containing Complex Metal Cyanides, in Relation to Concentrations of Molecular HCN." *Trans. Amer. Fish. Soc.*, **95**, **1**, 6 (1966).
3. Schneider, C. R., and Freund, H., "Determination of Low Level Hydrocyanic Acid in Solution Using Gas-Liquid Chromatography." *Anal. Chem.*, **34**, 69 (1962).
4. Claeys, R. R., and Freund, H., "Gas Chromatographic Separation of HCN on Porapak Q—Analysis of Trace Aqueous Solutions." *Environ. Sci. & Technol.*, **2**, 458 (1968).
5. Baker, B. B., and Morrison, J. D., "Determination of Microgram Quantities of Fluoride and Cyanide by Measurement of Current from Spontaneous Electrolysis." *Anal. Chem.*, **27**, 1306 (1955).
6. Strange, J. P., "Potentiometric Recorder for Hydrogen Sulfide and Hydrogen Cyanide." *Anal. Chem.*, **29**, 1878 (1957).
7. McCloskey, J. A., "Direct Amperometry of Cyanide at Extreme Dilution." *Anal. Chem.*, **33**, 1842 (1961).
8. Miller, G. W., *et al.*, "Submicromicrogram Determination of Cyanide by a Polarographic Method." *Anal. Chem.*, **36**, 980 (1964).

K. M. Grasshoff

K. M. Chan

An automatic method for the determination of hydrogen sulphide in natural waters

The most sensitive method for the determination of hydrogen sulphide is based on the reaction of N-1-dimethyl-*p*-phenylenediamine with hydrogen sulphide in the presence of iron(III) to form methylene blue[1]. A modified procedure has been applied to sea water by Fonselius[2], and the method has also been used for the determination of hydrogen sulphide in gases[3,4]. The formation of methylene blue is specific for hydrogen sulphide and has advantages over the usual iodimetric titration methods, which are time-consuming and not very sensitive, as well as inselective unless prior separation of metal sulphide is used. The methylene blue method is now applied as a standard procedure for the determination of hydrogen sulphide in sea water.

However, the methylene blue method is far too sensitive to be applied without modification for the entire range of concentration expected (0–500 μg at. H_2S–S per l). Previous experiments showed that Beer's law is only obeyed up to about 60 μg at. H_2S–S per l for the direct determination. Increasing the amount of reagents did not improve the linear range, hence a technique was developed which allowed dilution of the sample before the formation of the methylene blue without any loss of the sulphide by evaporation or oxidation. This could be done by immediate stabilization of the sampled sulphide as zinc or cadmium sulphide which was then kept in colloidal solution with gelatine. The analysis could then be continued either directly, or after proper dilution if the concentration exceeded 50 μg at. H_2S–S per l, *i.e.* if the absorbance exceeded 0.8 in a 1-cm cuvette. The initial formation of colloidal metal sulphide opened a way for the automation of the dye formation and the photometric determination by means of the Auto Analyzer technique.

Experimental

Sampling technique. The water sample for the determination of the hydrogen sulphide is taken with the same precautions as for the determination of oxygen[5]; 50–60-ml glass bottles with glass stoppers are sufficient. The neck of the bottle and the stopper should be shaped so that no air bubbles are trapped after stoppering. Immediately after the sampling (before stoppering), 1 ml of the zinc or cadmium chloride–gelatine solution is added to the bottom of the sample with a syringe. The bottle is then stoppered, after the water in the bottle neck has been replaced without trapping air, and shaken for about 30 sec. Stored in the dark, the colloid is stable for at least 24 h. In saline waters cadmium sulphide remains colloidal longer, whereas zinc sulphide tends to coagulate and precipitate. (Other colloid protectants may improve the stability of the colloidal sulphide.) The samples are then inserted into the Technicon sampler and diluted automatically to the proper concentration range.

Equipment. The Technicon AutoAnalyzer consists of the following modules: a sampler II, proportioning pumps with manifolds as shown in Fig. 1, colorimeter with a 660-nm filter and a 50-mm flow cell; a recorder and a range expander. The sample cycle is set at the rate of 20 per h.

Reagents. For the N-1-dimethyl-*p*-phenylenediamine dihydrochloride solution, dissolve 1 g of reagent in 500 ml of 6 *M* hydrochloric acid. Dilute 10 ml of the prepared stock solution to 120 ml with distilled water.

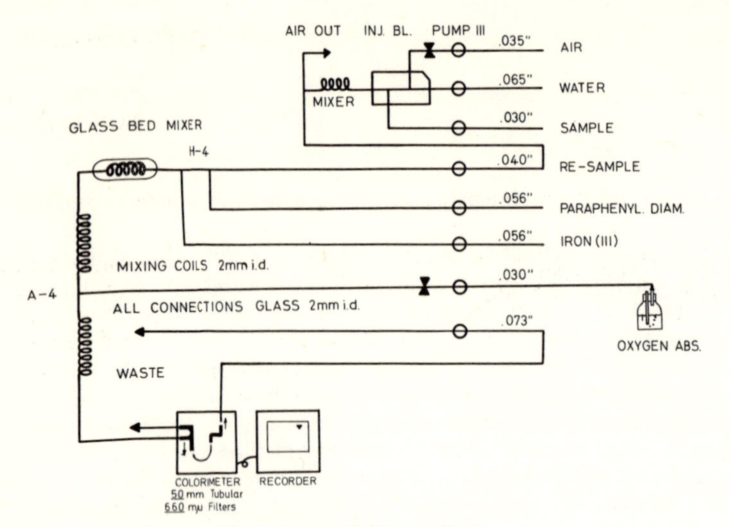

Fig. 1. Flow diagram of the hydrogen sulphide manifold.

For the iron(III) solution, dissolve 8 g of iron(III) chloride in 500 ml of 6 *M* hydrochloric acid. Dilute 10 ml of the stock solution to 120 ml with distilled water.

For the stabilizer solution, dissolve 7 g of zinc chloride in 400 ml of water, add 50 ml of 1% (w/v) gelatine solution and dilute to 500 ml.

Standard solution. Dissolve *ca.* 1 g of sodium sulphide nonahydrate (4 mg at. S^{2-}–S) in 100 ml of deoxygenated water (by boiling or by stripping with nitrogen). Use only large transparent crystals which have been washed with water and dried with filter paper. Transfer 10 ml of this solution to 1 l of a deoxygenated solution containing 0.2 g of zinc chloride and 2 g of gelatine. Stored in a brown glass bottle, this solution is stable for about 14 days.

To standardize the solution, pipette 50 ml into a 25-ml Erlenmeyer flask, and add 10 ml of a standard 0.0100 *N* potassium iodate solution and *ca.* 100 mg of potassium iodide. After dissolution, add 2 ml of 1 *M* sulphuric acid and, after 5 min, titrate the liberated iodine with standard 0.02 *N* thiosulphate solution. Carry out a blank titration on 100 ml of distilled water instead of the sulphide solution. Calculate the content of the sulphide solution from

$$\text{ml } S_2O_3^{2-} \cdot N \cdot 10^4 = C_{H_2S} \ (\mu\text{g at. } H_2S\text{–S l}^{-1})$$

(N = normality of the thiosulphate solution).

Pipette appropriate amounts of this diluted standard into 100-ml volumetric flasks and dilute to the mark with deoxygenated water (Fig. 2).

Results and discussion

The manifold. A modification of the AutoAnalyzer technique which normally involves immediate segmentation of the sample stream, must be applied for the determination of hydrogen sulphide. When the strongly acidic reagents are added, the colloidal metal sulphide is dissolved and the hydrogen sulphide tension in the liquid phase increases to very high values. If a gas phase were present (air segments), the hydrogen sulphide would tend to escape into the gas phase until equilibrium was reached. On the other hand, the concentration of the hydrogen sulphide in the liquid phase is diminished (to almost zero) by the formation of the methylene blue. The rate of the dye formation would then be controlled by the very slow re-entry of the hydrogen sulphide from the gas phase into the acid liquid phase. Accordingly, the sample stream is best segmented first when the methylene blue formation is nearly complete and there is almost no free hydrogen sulphide in the liquid phase.

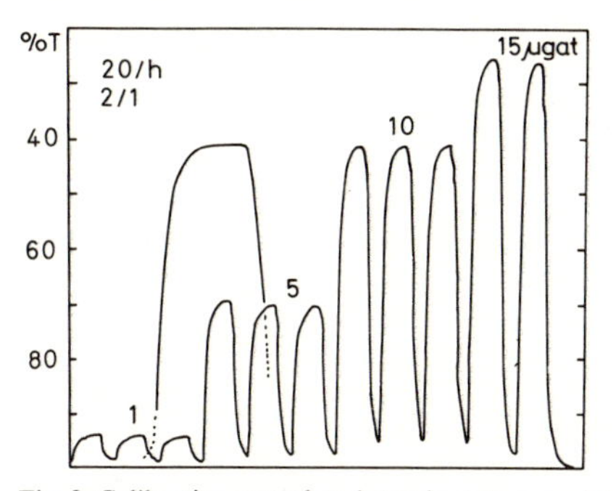

Fig. 2. Calibration record and steady state record for the concentration range 0–15 µg at. H_2S–S l^{-1}.

TABLE 1

LINEAR RANGES OF CALIBRATION CURVES WITH DIFFERENT TUBE SIZES

Sample tube (in.)	0.030	0.030	0.030	0.030
Dilution tube (in.)	0.081	0.100	0.100	0.100
Resample tube (in.)	—	0.081	0.030	0.040
Linear range (µg at. H_2S–S l^{-1})	0–50	0–100	0–200	0–300
Approximate absorbance at top of range	0.250	0.320	0.350	0.500

Beer's law. In the manual method, deviation from Beer's law at high concentration of hydrogen sulphide is observed. Similar results were also found when samples containing more than 30 µg at. H_2S–S l^{-1} were automatically analyzed without dilution. However, with a simple automatic dilution technique, as adopted in the present manifold, a linear calibration curve in the range of 0–300 µg at. H_2S–S l^{-1} can be achieved. With different combinations of tubing sizes, the manifold can be modified for any specific range required. Typical examples are shown in Table I.

Effect of acidity and temperature. The formation of methylene blue depends both on the temperature of the reaction and the acidity of the reagents. The reaction is faster at higher temperature and acidity, but the possibility of hydrogen sulphide escaping from the acid solution into the vapor phase before reacting increases and the reproducibility of the results is poor. Tests showed that the absorbance of the methylene blue solution was essentially constant when the acidity lay in the range 0.4–1 *M*. The results showed that the acidity of the reagent employed would give maximum color development without loss of hydrogen sulphide during the reaction.

Accuracy of the method. Replicate analyses of sea water samples containing sulphide of concentration of 1 μg at. H_2S–S per l with a 2 × range expander were done; the coefficient of variation of the results is ±0.8%. The sensitivity is 0.2 μg at. H_2S–S per l.

It seems possible with this method to collect water samples, stabilize the hydrogen sulphide and analyze the sulphide in a central laboratory.

We wish to express our sincere thanks to the Technicon Corporation for providing an AutoAnalyzer unit, and to the National Science Foundation and to the German Research Association (Grant Gr320/1) for financial support (Grant GA 1261).

1 E. Fischer, *Ber. deut. chem. Ges.*, (1883) 2234.
2 S. H. Fonselius, *Fish. Bd. Swed., Ser. Hydrogr. Rep.*, 13 (1962) 31.
3 J. K. Fogo and M. Popowsky, *Anal. Chem.*, 21 (1948) 732.
4 W. Sonnenschein and K. Schäfer, *Z. Anal. Chem.*, 140 (1953) 16.
5 K. Grasshoff, *Kiel. Meeresforsch.*, 18 (1962) 42.

THE RESPONSE OF AQUATIC COMMUNITIES TO SPILLS OF HAZARDOUS MATERIALS

JOHN CAIRNS, JR.
KENNETH L. DICKSON
JOHN S. CROSSMAN

Aquatic ecosystems have the ability to assimilate a certain amount of waste material and maintain near normal function. With the constant use and reuse of water by industries, agriculture and municipalities, the function may be altered or disrupted if the assimilative capacity is exceeded. The ability of a river or lake to assimilate wastes is governed by the capacity of the system to transform them before they reach deleterious levels. If an overload occurs due to a spill of a hazardous material, the system is disrupted and the transforming capacity may be substantially reduced.

The capacity of a system to receive and assimilate spills of hazardous materials without environmental damage is determined by the physical, chemical and biological conditions of the receiving system. Introduced materials may be transported, rendered, converted, respired, incorporated, excreted, deposited and thus assimilated by the system. However, not all systems can receive and assimilate the same quantity or kinds of waste material and, even within a single system, this capacity will vary. The capacity of each system to resist shock loadings of wastes without significant damage is a function of complex environmental factors. Physical factors in a river or stream such as flow velocity, volume of water, bottom contour, rate of water exchange, currents, depth, light penetration, temperature, etc., as well as the biological factors, govern in part the ability of a system to receive and assimilate waste or spills of hazardous materials.

The unique chemical characteristics of different systems govern in part the kinds and quantity of wastes a system can receive without damage. Due to synergistic or antagonistic interactions with receiving chemical water quality, the effects of a potentially hazardous material on a wide variety of receiving systems is difficult to predict.* The modification of the toxicity of potentially hazardous materials by chemical characteristics such as, hardness, alkalinity, pH, buffering capacity and nutrients may markedly affect the ecological effects.

The biological community present in a receiving system is determined partly by the physical and chemical characteristics of the system and partly by the kinds of organisms that colonize the area (this latter is often a matter of random chance). Obviously, if systems vary chemically and physically, they then vary in the components of the biological community present. Thus the effects of a potentially hazardous material introduced through a spill may vary from one system to another, because not all organisms will respond in the same way to a particular concentration of a hazardous material. However, in the past twenty years a great deal of evidence has accumulated which indicates that biological communities are not a haphazard collection of species,

[1]There are some well regarded water pollution specialists in Great Britain who believe this is possible, and they have evidence supporting their point of view (Herbert and Van Dyke, 1964; Jordan and Lloyd, 1964; Brown *et al*, 1965).

thrown together by the whims of nature, but rather that they have a fairly well defined structure. While the individual components of the aquatic community may vary from one aquatic system to another, the structure of the communities are remarkably similar in most temperate zone fresh-water systems in industrialized areas.

Essentially this structure is due to the fact that energy flows through a diverse network of species called a food web (Figure 1). Although the number

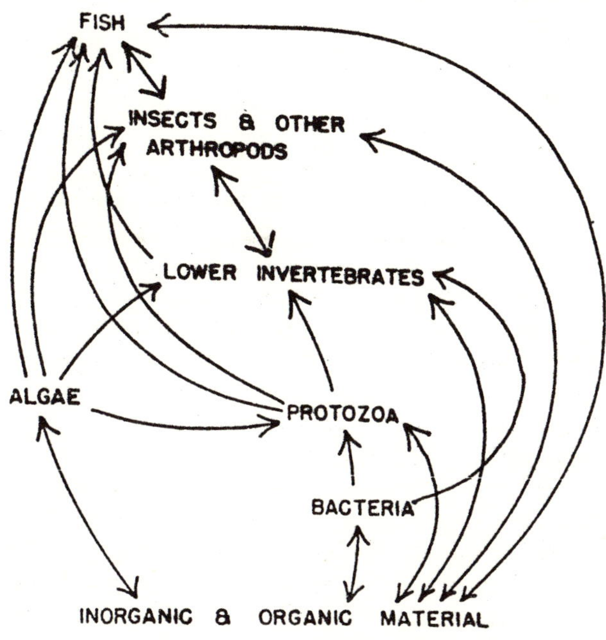

Figure 1: An Aquatic Food Web.

of individuals at any one energy level may vary, the number of species involved in the web (or to put it another way, the number of qualitative differences possible in the system), remains relatively constant under natural conditions. The number of components of species in this energy system also remains remarkably constant from one river basin to the next, and at various points within a basin as well, despite the fact that the types of species which comprise the system may differ at various sampling points. This is merely a restatement of the survival of the fittest principle with the corollary assumption that there are several species available to do each particular job, and that one of these may outcompete its rivals in one river system, but not in all river systems. Recognition of this principle obviates the need for a detailed study of a miscellaneous collection of aquatic organisms in each receiving stream, and restricts the problem to the study of basic changes in the system itself.

Response of Aquatic Communities to Stress

In any population consisting of various ages, sizes, sexes, etc. of a single species, there will be some individuals that are more sensitive to a particular type of stress than the others. This is true whether the population consists of humans or fish. There is usually a range of concentrations, temperatures, etc. which the organisms will tolerate without any adverse effects. It is important to note that this is true for some very toxic materials such as cyanide, and that the reverse is true for most beneficial substances such as table salt or vitamin A (i.e. unfavorable and even lethal responses). When the concentration or application of the stress increases beyond the point tolerated by the organism, the more sensitive or weaker individuals in the population will begin to respond (in the case of the lethal threshold, by dying). Gradually as the stress increases, more and more members of the population will respond until the entire population is doing so. The important thing is that all organisms, even of a single species, do not respond in the same way to the same level of stress. In practical terms this means that an industrial spill that causes substantial numbers of the population to respond is probably well above the response concentration for the weaker members, and well into the concentration causing a response for the stronger members. Thus nearly the entire population is affected. In general, the amount of stress that affects only a relatively small number of organisms in a population goes unnoticed unless an experienced biologist happens to be gathering data on the situation in question.

The aquatic community which consists of a mixture of species follows essentially the same pattern, that is, the species most sensitive to the stress (which are generally those species living under marginal or suboptimal conditions) disappear while the more resistant individuals may do even better and increase in numbers. This is shown in Figures 2 and 3; notice that the curve in Figure 3 is depressed when one compares it to the curve in Figure 2, but the tail of the curve in Figure 3 is much longer, indicating that a few of the species have done even better than those in Figure 2 once their competitors were partially removed. This is an important characteristic of the response of complex communities because it shows

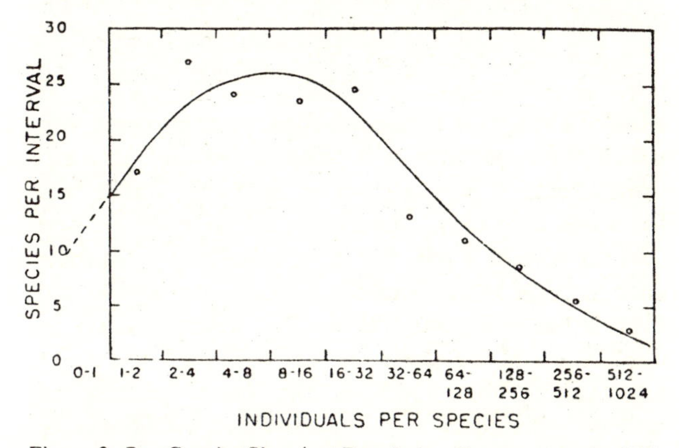

Figure 2: Bar Graphs Showing Population Structure under Different Degree of Pollutional Stress.

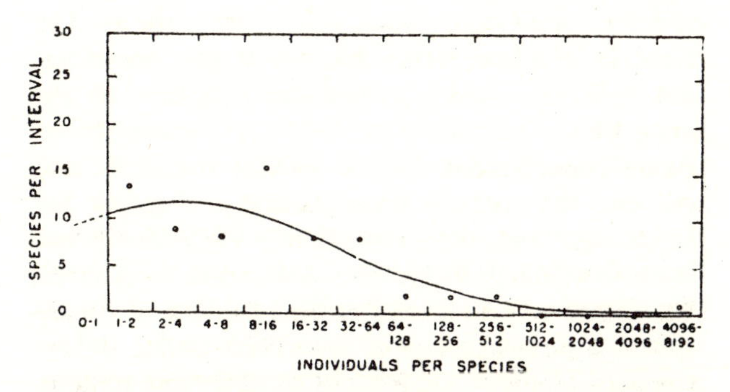

Figure 3: Graph of Diatom Population from a Section of Ridley Creek, Chester County, Pennsylvania, not adversely affected by pollution.

that the initial response to stress may be a change in the complexity of the system (i.e. the number of species), but a compensation mechanism in numbers of individuals may result in about the same biomass being present in each of the two communities. However, the biomass is represented by few species in the stressed community. This means that one has more "eggs in one basket," and any natural event which eliminates one of the high density species in the stressed community is going to have a much more significant effect than the same stress eliminating the same species in an unstressed community. Thus the complex communities are more stable in their performance because complex communities are less affected by the continual replacement of species which is characteristic of the natural successional process.

Most of the species in natural communities are present in relatively low numbers because the competition is keen. However, when pollutional stress removes some of the competition, the survivors become overabundant and then are termed pest species. The dieoffs of one of these overabundant species, even when they consist of algae or other lower life forms, can cause taste and odor problems, clogged filters, lower the oxygen content of the water and otherwise disrupt an aquatic system. The loss of a single species in the diverse low number of individuals per species system, however, is not nearly so disruptive. This, of course, is a rather crude and grossly oversimplified statement of the "balance of nature".

Case History Studies

The response of aquatic communities to spills of hazardous materials can be divided into the immediate response to acute stress; the response to chronic stress from materials with residual toxicity and the recovery response of the community following removal of the stress. The following case history studies of ethyl benzene-creosote, acid, and caustic spills illustrate the response of the various components of the aquatic community to drastically altered environmental conditions. The natural biological recovery following acute stress resulting from the spill of hazardous materials is documented and contrasted in these case history studies.

The Biological Damage and Recovery from an Ethyl Benzene-Creosote Spill

The effect of an abrupt release of acutely toxic material into a river is an immediate stress on the organisms in the river, followed by dilution of the waste and eventual restoration of water quality. A spill of this nature occurred October 10, 1970, at Salem, Virginia, on the Roanoke River. Ethyl benzene mixed with creosote was spilled into the Roanoke River from a primary storage tank of the Koppers Company. Approximately 2,000 gallons of solvent escaped and entered an open cooling water ditch (100 gal/min) which flowed into the Roanoke River. Approximately 400 to 600 gallons of the solvent entered the river in a period of 1 to 2 hours. River flow at the time of the spill was 19,000 gal/min, resulting in an estimated concentration of solvent at the point of discharge of 1,000 ppm. The subsequent fish kill was reported by the Virginia Water Pollution Control Board, and a total of 13,281 fish were killed consisting of 7,979 rough fish and 5,302 sport fish. Initial estimates by the water control board indicated that biological damage extended for a distance of 7 miles below the plant's outfall.

The exact toxicity of the ethyl benzene-creosote mixture is not known. However, using static bioassays, Cairns and Scheier (1959) found that the 96 hour TL_m for *Lepomis macrochirus* (bluegill sunfish) was 10.0 ppm for creosol. Pickering and Henderson (1966) established the 96-hour TL_m for the bluegill using ethyl benzene at 29 ppm under static conditions. The synergistic or antagonistic actions of these compounds, or their action with other compounds in the river could have altered the acute toxicity of the spilled material in the Roanoke River.

Sampling stations were selected above and below the site of the spill and were sampled for both fish and aquatic macroinvertebrates. Sampling was done between 6 and 10 days after the spill for aquatic invertebrates, and 9 to 11 days after the spill for fish. A follow-up bottom fauna and cursory fish survey was conducted April 1 and 2, 1971, (six months later) to determine the extent of recovery. All sites were selected for their ecological similarity, having shallow riffles with rock and gravel beds with heavily wooded banks. The sampling stations were as follows:

Reference Station 1—Located 50 yards above Virginia Rt. 612 Bridge and one-half mile upstream from the spill site.

Reference Station 2—Located adjacent to Koppers Plant and .03 miles upstream from the spill site.

Station 3—Located 0.5 miles below the site of the spill with access via Krogers Road.

Station 4—Located approximately 2.0 miles below the site of the spill with access via Duguid Road.

Station 5—Located 3.5 miles below Koppers with access via Mill Street.

Station 6—Located approximately 4.5 miles below Koppers with access via Union Street.

Station 7—Located approximately 6 miles below Koppers with access via Colorado Street.

Fish Survey

Fish were collected using a 10 foot x 4 foot seine of 1/4 inch mesh. All ecologically distinct sections of the stream were searched for fish at each of the seven sampling stations.

Jordan published a list of fishes taken from several places in and near Salem, Virginia (1890: 120-124). Since then several collectors have sampled the river, including Raney and Ross. In 1952 they made a large and representative collection of the Roanoke River at the Route 11 crossing, on the Montgomery-Roanoke County line (V. P. I. & S. U. #506). There were 33 species in the collection, now preserved in the Cornell University Fish Collection at Ithaca, New York (Table 1). Jordan's results listed 28 species (Table 1). Four of the species obtained by

Raney and Ross in 1952 were not recognized by Jordan, those marked with an asterisk. These collections are closely comparable and represented the former stream faunistic composition in the area under study.

Based on the survey of the Roanoke River conducted immediately after the spill, the fish fauna appeared to be in a state of good health except for the portions of the river delimited by Stations 3 and 4 (Figure 4). A total of 26 species of fish were found at Station 1 with 20 species at Reference Station 2. Smallmouth bass, bluegills, and rock bass were present at the reference stations, but were absent at Stations 3 and 4 downstream (Table 2). Six species of perches were present at Reference Station 1, but were absent at Reference Station 2. Their absence was probably due to the introduction of sediments from highway and construction projects located between the two stations. Stations 3 and 4 below the site of the spill had a paucity of fish with only four species of fish (all minnows) being found at Station 3 and eight at Station 4 (Table 2).

At Station 5 approximately three and one-half miles below the Koppers Plant, 23 different species of fish were found, making it comparable to the diversities found at the reference stations. Stations 6 and 7 were highly productive with diversities similar to those found at the reference stations. They also supported a greater biomass of fish than noted at the other stations. This appeared to be due to nutrient enrichment which resulted in stimulated algal growth.

Bottom Fauna Survey

Four Surber Square Foot Samples were taken in a transect across the Roanoke River at each sampling station to quantitatively determine the effects of the spill on the bottom fauna. A bottom net sample was also taken at each station for qualitative evaluation of species diversity. Bottom organisms were separated from the debris using a number 30 mesh sieve and preserved in 70% ethanol for enumeration and identification.

Based on the survey of the Roanoke River conducted October 19-21, 1970, the bottom fauna at Stations 3 and 4 were drastically reduced when compared to Reference Stations 1 and 2. Eighteen taxa of bottom fauna were found at Reference Station 1 with 20 found at Reference Station 2 (Figure 4).

Construction of a sewer line introduced suspended solids between the two reference stations and may have been responsible for the decrease in density at Reference Station 2 (Figure 5). Mayflies, stoneflies, and caddisflies (pollution intolerant organisms) were present at the reference stations, but were absent at Stations 3 and 4 (Table 3).

Stations 3 and 4 located below the site of the spill had a typical "pollution tolerant" community. Pollution tolerant snails and midge larvae made up the major part of the community. Fourteen taxa of bottom fauna were found at Station 3 with 17 found at Station 4. The average number of organisms/ft^2 was lower at these stations when compared to Reference Station 1, probably indicating an additive effect of sedimentation and toxicity from the spill.

Urbanization and the introduction of other industrial discharges into the Roanoke River influenced the character of the benthic communities at Stations 5, 6, and 7. *Cladophara*, a green algae was present at Stations 5, 6, and 7 indicating nutrient enrichment. Twenty taxa of bottom fauna were present at Station 5 with extremely numerous herbivorous caddisfly larvae and snails contributing to the high productivity of the area. Stations 6 and 7 supported a bottom fauna community typical of a nutrient enriched area.

A numerical diversity index evaluation ($\bar{d}$) (Wilhm and Dorris, 1968) indicated that Stations 1, 2, and 4 were healthy stations. All other stations qualified as mildly polluted (Figure 6). However, based on density, diversity, taxonomic information and the diversity index evaluations, it appeared that the degradation of the Roanoke River due to the Koppers Company operations and/or spill was restricted to that area extending not more than three and one half miles below the plant.

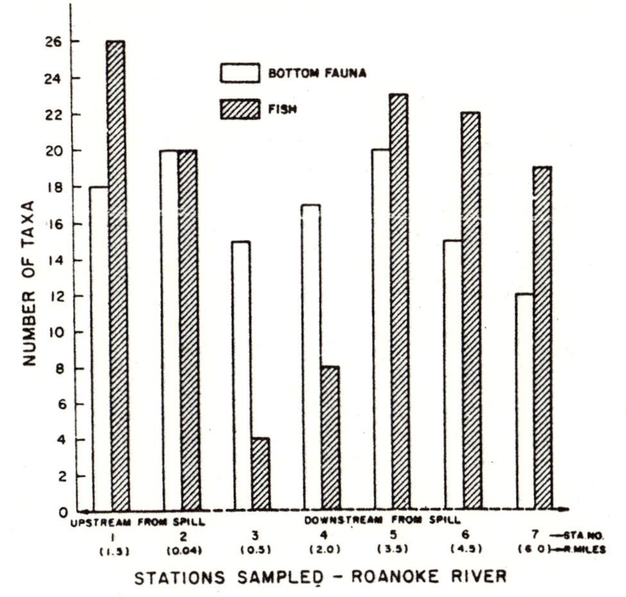

Figure 4: The Number of Bottom Fauna and Fish Found at Each Sampling Station.

Recovery Survey

A fish and bottom fauna survey was conducted in early April (approximately six months after the ethyl benzene-creosote spill) to document the extent of recovery of the damaged areas.

The results of the bottom fauna survey showed that riffle beetles which were present at Station 3, 6-10 days after the spill were absent six months later, possibly due to either a residual toxicity from the insoluble creosote or from continued introduction of suspended solids from construction (Table 3). Mayflies and stoneflies had recolonized the affected areas, indicating fairly good water quality since these organisms are generally considered "pollution intolerant" organisms.

TABLE 1

Fishes Taken by Jordan in 1888 and by Raney and Ross in 1952
from the upper Roanoke River in Montgomery and Roanoke Counties, Virginia

	Jordan 1888	Raney & Ross 1952
Suckers, family Catostomidae.		
Hog sucker, *Hypentelium nigricans*	common	common
Roanoke hog sucker, *H. roanokense*		common*
Rustyside sucker, *Thoburnia rhothoeca*		several*
Common sucker, *Catostomus c. commersoni*	common	common
Black jumprock, *Moxostoma cervinum*	common	common
Bigeye jumprock, *M. ariommum*		common*
V-lip redhorse, *M. collapsum*		common
Suckermouth redhorse, *M. papillosum*	few	one
Minnows, family Cyprinidae.		
Rosyside dace, *Richardsonius funduloides*	scarce	
Bluehead chub, *Nocomis leptocephalus*	abundant	common
Nocomis sp.		several*
Blacknose dace, *Rhinichthys a. atratulus*	common	scarce
Cutlips minnow, *Exoglossum maxillingua*	common	common
Bluntnose minnow; *Pimephales notatus*		
Silvery minnow, *Hybognathus nuchalis regius*		one
Stoneroller minnow, *Campostoma anomalum michauxi*	common	several
Mountain redbelly dace, *Chrosomus oreas*	abundant	several
Mimic shiner, *Notropis volucellus*		common
Rosefin shiner, *Notropis a. ardens*	common	abundant
Crescent shiner, *Notropis cerasinus*	abundant	abundant
White shiner, *Notropis cornutus albeolus*	common	common

Satinfin minnow, *Notropis analostanus*	common	
Swallowtail shiner, *Notropis procne longiceps*	not rare	common
Spottail shiner, *Notropis hudsonius saludanus*		one
Catfishes, family Ictaluridae.		
Orangefin madtom, *Noturus gilberti*	present	several
Margined madtom, *Noturus insignis*	common	common
Sunfishes, family Centrarchidae.		
Smallmouth bass, *Micropterus d. dolomieui*	scarce	several
Redbreast sunfish, *Lepomis auritus*	common	common
Pumpkinseed, *Lepomis gibbosus*	scarce	
Roanoke rockbass, *Ambloplites rupestris cavifrons*	two	
Rockbass intergrades *A. r. rupestris x A. r. cavifrons*		several
Perches, family Percidae.		
Roanoke logperch, *Percina rex.*	two	few
Piedmont darter, *Percina crassa roanoka*	common	common
Shield darter, *Percina peltata*	several	several
Fantail darter, *Etheostoma f. flabellare*	common	common
Riverweed darter, *Etheostoma podostemone*	common	common
Johnny darter, *Etheostoma nigrum nigrum*	one	several
Eels, family Anguillidae.		
Common eel, *Anguilla rostrata*	common	

*Species not recognized by Jordan.

A cursory fish survey (six months after the spill) of the upstream and downstream area around the Koppers Plant indicated that:

(1) The upstream area (Reference Station 1) still supported a healthy and diversified fish fauna. Approximately 28 species of fish were present including representatives from the sucker, minnow, catfish, sunfish, and perch families.

(2) The area immediately below the site of the spill supported a suppressed population of fish generally represented by the minnows. Apparently the minnows can reinvade more rapidly or are more resistant to any lingering toxicity or discharges from the Koppers Plant than are the other groups of fish.

SUMMARY

(1) The effects of an acute stress from an industrial spill of ethyl benzene and creosote were to decrease the diversity and density of both the fish and bottom fauna for approximately 3 miles below the site of the spill.

(2) A differential response to the stress was apparent with all major groups of fish except the minnows being entirely eliminated in the stressed area. Perhaps the minnows avoided the stress by swimming into small isolated tributaries or perhaps they rapidly reinvaded the area, or they may be more tolerant to this type of stress.

(3) Mayflies, stoneflies, caddisflies, and mussels did not survive the stress and were eliminated. However, mayflies and stoneflies were present six months later indicating improved water quality.

(4) Riffle beetles, trueflies, crayfish, some snails, and segmented worms were apparently able to survive the short term exposure to the ethyl benzene and creosote stress.

The Biological Damage and Recovery of the Clinch River Following Acute pH Stresses

The Clinch River has been subjected to two major industrial spills which have resulted in fish kills and elimination of other aquatic organisms. To evaluate the effects of these spills and to study the recovery processes, benthic organisms and fish have been collected from the Clinch in southwestern Virginia and northeastern Tennessee for two years.

Description of the Clinch River

The Clinch River is a headwater tributary of the Tennessee River located in the ridge and valley region of southwestern Virginia and eastern Tennessee. As the main tributary of the Clinch River Basin, the Clinch drains 1,260 square miles of land in Virginia and extends for a distance of 148 river miles before it enters Tennessee (Tackett, 1963). The average discharge for the Clinch at Speer's Ferry, Virginia, during the period of 1920 to 1960 was 1,578 c.f.s. with a maximum of 45,300 c.f.s. noted in January, 1957, and a minimum of 42 c.f.s. noted in September, 1939 (U.S. Geological Survey, 1960).

As the Clinch passes through the mountainous regions of Virginia it flows over exposed geologic formations of limestone and dolomite which range in age from early Cambrian to Pennsylvanian (Cooper, 1945). These formations contribute calcium bicarbonate to the water and raise the pH of the river to 8.0–8.5. Because of the high alkalinity, acid discharges from coal mines are readily neutralized. However, discharges from coal washing operations cause a major sedimentation problem in the Guest River and Dump's Creek, two of the larger tributaries of the Clinch.

1967 Fly Ash Pond Spill

The first fish kill occurred when the dike surrounding a fly ash holding pond collapsed at Appalachian Power Company's 700-megawatt steam power generating plant near Carbo, Virginia. At this power plant native coal is utilized to produce steam which is used in the production of electrical power. Because the coal has a high ash content, approximately 960 tons of fly ash is produced daily. To efficiently remove such large quantities of ash from the furnace hoppers, water from the Clinch River is mixed with the ash to form a slurry. This

TABLE 2

The numbers of fishes collected at 7 stations (pagell) from the Roanoke River on October 19-21, 1970 in Roanoke County and Salem, Virginia.

Stations:	1	2	3	4	5	6	7
Suckers, family Catostomidae.							
Hog sucker, *Hypentelium nigricans*	3	20		2	7	11	14
Rustyside sucker, *Thoburnia Hamiltoni*	12	10		2	1	14	7
Common sucker, *Catostomus c. commersoni*		1			1		
Black jumprock, *Moxostoma cervinum*	16	12		1		1	18
Bigeye jumprock, *Moxostoma ariommum*	2	8			2	6	6
V-lip redhorse, *Moxostoma collapsum*	4	4			6	1	2
Suckermouth redhorse, *Moxostoma papillosum*		1			1		
Minnows, family Cyprinidae.							
Bluntnose minnow, *Pimephales notatus*	36	15			3	4	
Bluehead chub, *Nocomis leptocephalus*	10	13	2	4	4	21	20
Nocomis sp.		8				10	35
Cutlips minnow, *Exoglossum maxillingua*	1						
Silvery minnow, *Hybognathus nuchalis regius*	1						
Stoneroller minnow, *Campostoma anomalum michauxi*	15	69			10	54	42
Mountain redbelly dace, *Chrosomus oreas*					1		
Rosefin shiner, *Notropis a. ardens*	86	151	3	15	8	144	142
Crescent shiner, *Notropis cerasinus*	126	77	19	47	15	64	29
White shiner, *Notropis cornutus albeolus*	125	8	4	46	109	50	76
Satinfin shiner, *Notropis analostanus*	1	1			1	7	1
Swallowtail shiner, *Notropis procne longiceps*						35	3
Spottail shiner, *Notropis hudsonius saludanus*		1			4		

Catfishes, family Ictaluridae.							
Orangefin madtom, *Noturus gilberti*	1						
Margined madtom, *Noturus insignis*	17	24		1	8	8	16
Sunfishes, family Centrarchidae.							
Smallmouth bass, *Micropterus d. dolomieui*	2				2		
Bluegill, *Lepomis m. macrochirus*		2			2		
Redbreast sunfish, *Lepomis auritus*	7	2			3	6	
Pumpkinseed, *Lepomis gibbosus*	1						
Rockbass intergrades, *Ambloplites r. rupestris x A. r. cavifrons*	1	2			1		1
Perches, family Percidae.							
Roanoke logperch, *Percina rex*	15				1	1	2
Piedmont darter, *Percina crassa*	58				35	88	61
Shield darter, *Percina peltata*	1						
Fantail darter, *Etheostoma f. flabellare*	7				1	66	12
Riverweed darter, *Etheostoma podostemone*	7				7	34	18
Johnny darter, *Etheostoma n. nigrum*	2					1	

mixture is pumped to large settling lagoons where the ash settles and the supernatant is recycled. Because of recycling, free lime (CaO) in the fly ash reacts with water to form $Ca(OH)_2$. This gradually raises the pH of the recirculating water and the water in the fly ash lagoons to extremely high pH values ranging from 12.0 to 12.7 (Anonymous, 1967b).

In June 1967, a 50–75 foot section of a dike surrounding one of the fly ash settling lagoons failed. Within less than an hour 400 acre feet (approximately 130 million gallons) poured into Dump's Creek, which joins the Clinch River 0.5 miles downstream. This caustic slug equalled 40% of the daily flow of the Clinch at the time and resulted in blocking the normal flow for several minutes. It also raised the water level several feet and forced some of the waste approximately 0.5 miles upstream.

For four and one-half days following the spill the alkaline slug traveled downstream at a rate of approximately 0.85 miles an hour killing essentially all the fish in its path (Anonymous, 1967b). During this period 162,000 sport and rough fish were killed in 66 miles of the Clinch River in Virginia. An additional 54,600 sport and rough fish were killed in 24 river miles in Tennessee until the polluted mass was diluted, dispersed, and neutralized in the river by natural physical-chemical forces. The chemical reactions involved in neutralization were:

1) $Ca(OH)_2 + Ca(HCO_3)_2 \longrightarrow 2CaCO_3 + 2H_2O$

In addition CO_2 from the atmosphere reacted with the water to form:

2) $H^2O + CO_2 \rightleftharpoons H_2CO_3$

3) $Ca(OH)_2 + H_2CO_3 \longrightarrow CaCO_3 + 2H_2O$

The lethal agent was reported to have been the high pH of the alkaline water which was composed of 90% hydroxide alkalinity and 10% carbonate alkalinity. A secondary effect which may have contributed to the biological damage was a depression in the dissolved oxygen concentration caused by the decaying organic matter (Anonymous, 1967b).

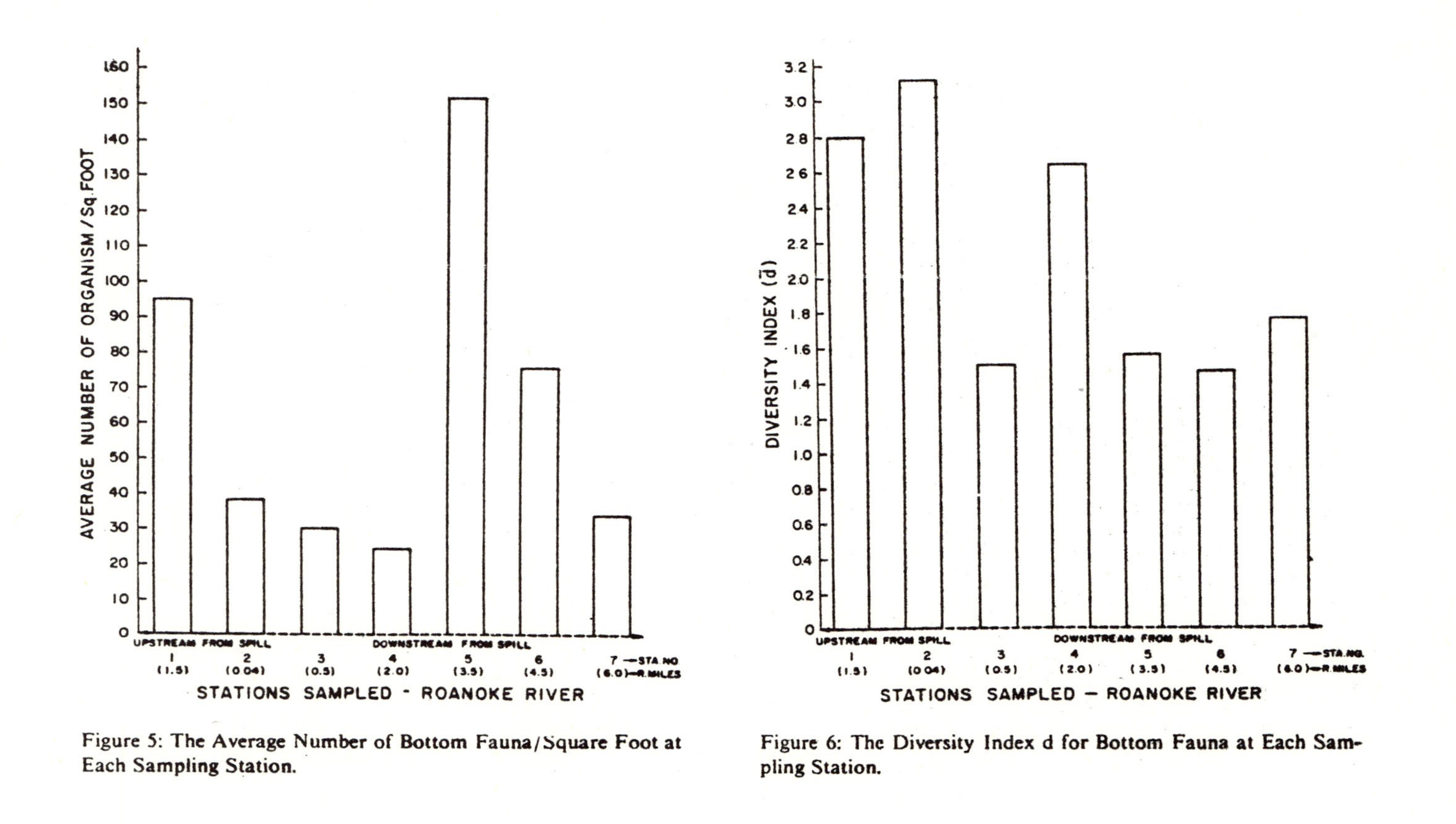

Figure 5: The Average Number of Bottom Fauna/Square Foot at Each Sampling Station.

Figure 6: The Diversity Index d for Bottom Fauna at Each Sampling Station.

TABLE 3

Number of Bottom Fauna Genera in Each Family by Stations

Bottom Fauna	Station No.								
	Six Days After Spill							Six Months After Spill	
	1	2	3	4	5	6	7	1	3
	Upstream		*Downstream*					*Upstream*	*Downstream*
Bettles	4*	3	3	2	4	2	2	2	0
Caddisflies	3	3	0	2	2	1	1	2	0
Damselflies	0	0	1	0	0	0	1	0	0
Dobsonflies	1	1	1	2	2	2	1	1	1
Mayflies	3	5	0	3	2	2	3	5	5
Stoneflies	1	1	0	1	1	0	0	2	2
Trueflies	4	3	2	3	4	2	1	1	2
Crayfish	1	1	1	1	1	0	1	1	0
Scuds	0	0	0	0	0	0	1	0	0
Sowbugs	0	0	0	1	0	0	0	0	0
Flatworms	0	0	1	0	0	0	0	0	0
Mussels	1	1	0	0	1	0	0	0	0
Snails	1	1	3	2	2	2	1	1	1
Worms	1	1	1	1	1	1	1	0	0
Total Number	20	20	13	18	20	12	13	15	11

*Number of Genera

Ten days after the collapse of the dike the Virginia State Water Control Board conducted a bottom fauna survey at selected stations above and below the site of the spill to assess damage to the benthic fish food organisms (Anonymous, 1967a). They observed that:

1) Bottom dwelling fish food organisms appeared to have been completely eliminated for a distance of approximately 3 or 4 miles below the site of the spill (Figures 7 & 8).

2) A drastic reduction in the number and kinds of bottom dwelling fish food organisms occurred in the Clinch River for 77 miles below the spill (Figures 7 & 8).

3) Snails and mussels were eliminated for 11.5 miles below Carbo, Virginia.

4) The Virginia State Water Control Board, based on past experience, predicted that as far as total weight per square foot of stream bottom, the Clinch River would recover to its former productive capacity within three months after the spill. Both the Virginia and Tennessee Game Commissions believed that the stream organisms would appear in sufficient numbers for fish restocking in the fall of 1967 (Anonymous, 1967a).

Materials and Methods

During the summer and fall of 1969, a bottom fauna and fish survey of the Clinch River was conducted to study the extent of the biological recovery of the river following the 1967 fly ash spill. Particular emphasis was placed on studying the bottom fauna communities above and below the spill site because:

1) Benthic organisms are relatively sessile organisms and they cannot quickly avoid environmental stresses as fish often are able to do;

2) They have rather long and complex life histories and their presence or absence reflects the history of the environment;

3) Since they are members of the food web in an aquatic environment, their presence or absence directly effects fish populations;

4) Sampling techniques for bottom fauna are more reliable than techniques for fish; and

5) More biological information can be gained from studying this group of organisms per dollar invested than any other group.

The bottom fauna survey involved the locating and sampling of twenty-one ecologically similar stations which extended from Blackford, Virginia, to Sneedville, Tennessee, a distance of 120 river miles (Figure 9). Four stations were located above the site of the spill to serve as reference or control stations with which downstream stations could be compared. Twelve stations were located below the spill site to assess the river's recovery and to evaluate any contributing influences arising from industries, municipalities, or agricultural areas. Five additional stations were established on tributaries with pollution sources which could have effected the main river stations.

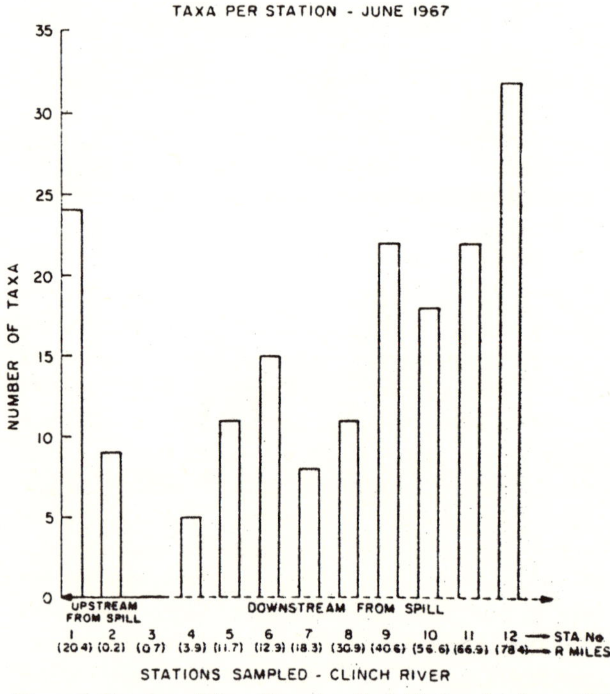

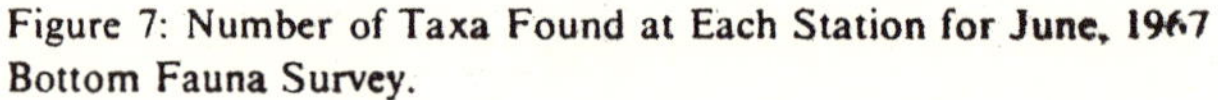
Figure 7: Number of Taxa Found at Each Station for June, 1967 Bottom Fauna Survey.

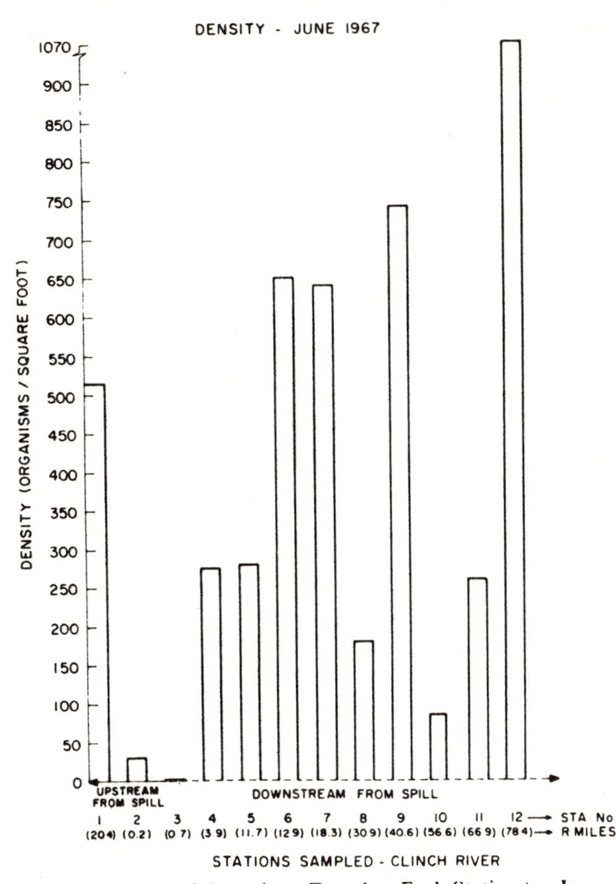

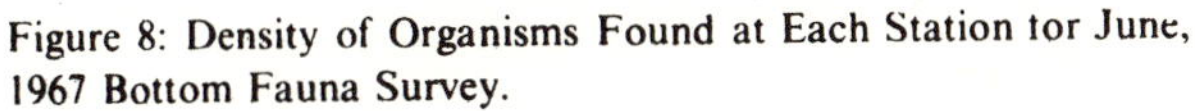
Figure 8: Density of Organisms Found at Each Station for June, 1967 Bottom Fauna Survey.

Sampling sites with comparable habitats and ecological similarity were selected so comparisons could be made between stations. After locating the stations, sampling was accomplished using a Turtox 8″ × 10″ × 18″ rectangular bottom net and a Surber square foot sampler. Care was taken that an equal collecting effort was made at each station. Each station was divided into three substations (left bank, right bank, and midchannel), and samples were taken from each substation with a bottom net since it had been observed that waste discharges were often restricted to certain portions of the river after discharge. Five Surber samples were taken along a transect through the riffle area for quanti-

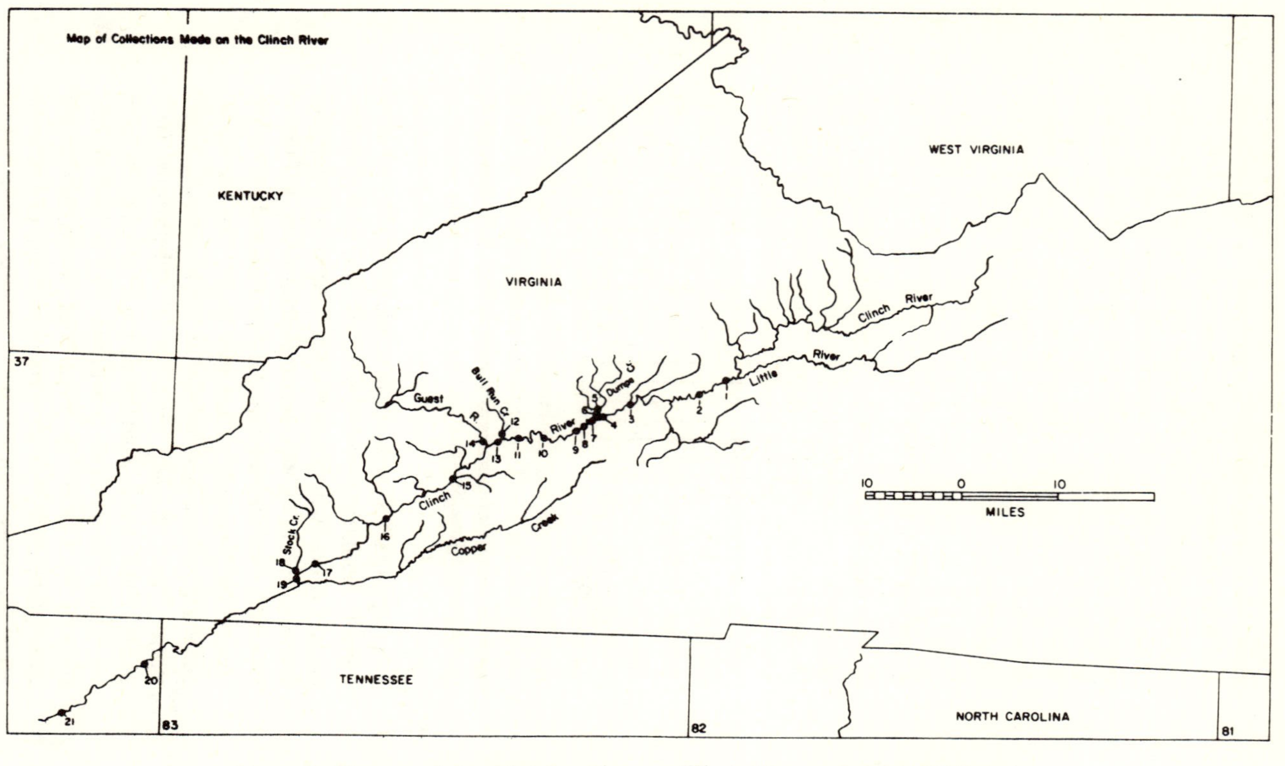

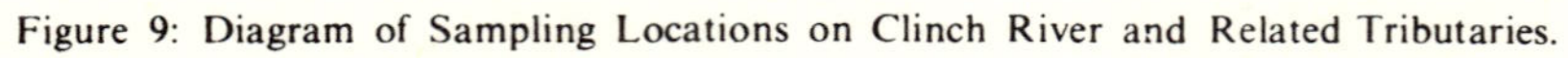

Figure 9: Diagram of Sampling Locations on Clinch River and Related Tributaries.

tative information. Immediately after collection, all samples were passed through a series of graded sieves (Tyler mesh Nos. 4, 10, and 35) and the organisms were removed and preserved in 70% ethanol for later identification and enumeration.

Fish samples were collected using a 10′ × 4′ seine of 1/4 inch mesh. All specimens were preserved in formalin and stored in 70% alcohol in the Virginia Polytechnic Institute and State University museum. The collections were obtained at two stations above the spill site and at four stations below. Only the shallow areas and riffles were sampled.

Results and Discussion

The number of bottom fauna taxa found at each station are summarized in a histogram, Figure 10. A difference existed between upstream reference stations and stations immediately below the spill site for a distance of approximately 18 miles. At Reference Stations 1 through 4 the number of taxa varied between 48 and 54, while below the spill site there was a decrease at Station 7 to 43 taxa followed by an even more pronounced decrease at Station 8 to 33. This low value was followed by an increase in the number of species at the next three stations until at Station 11 there were 46 taxa, approximately the same number as were observed at the reference stations. At Station 13 a decrease in the number of species was found, followed by an increased number of species equal to reference station values at the remaining downstream stations.

In Figure 11, a graph of density which includes all snail and freshwater mussel species, the difference between the upstream reference stations and stations below the spill site was even more pronounced. Densities for the reference stations ranged between 145.2 and 24.6 organisms per square foot. In contrast, densities at Stations 8-11 varied between 2.4 and 18.8 organisms per square foot with the highest density being found at Station 11, the station located furthest downstream. Station 13 showed a reduction in density which appeared to be due to unsuitable substrate conditions for benthic colonization and the influence of the tributary that flows into the Clinch

River just above this station.

Figure 12, a graph of density which does not include snails and mussel species, indicated a difference between Reference Stations 1-4 and Stations 7-11. In addition, Figure 12 also indicated that:

1) There was a reduction in the densities at Stations 1-4 and 13-21 when compared to the same stations in Figure 4. This was caused by the exclusion of the snail and mussel data, since these groups made up large portions of the invertebrate community at each station.

2) Densities were not affected by excluding snail and mussel species at Stations 7-11 because molluscs had not become reestablished after the spill. This may be the result of reduced susceptibility to downstream drift when compared to the drift found for insect larvae and other invertebrates, longer life cycles than found for most other aquatic invertebrates, and the lack of an aerial stage in their life cycles, all of which would affect their rate and extent of recolonization.

The differences found in diversities and densities at Stations 7-11 indicate a possible recovery pattern. This pattern would involve a combination of interrelated factors which would include the amount of damage at each station from the spill, differing rates of recolonization by different organisms, and a continuing mild environmental stress upon a limited portion of the aquatic ecosystem being studied.

The continuing environmental stress appears to have been caused by a discharge from the Appalachian Power Company's plant above Station 7. This discharge channeled along the right bank for some distance before mixing, and although it did not affect the overall diversity found at Station 7, it did appear to have an adverse effect upon aquatic organisms at downstream stations.

Community structure analyses of the bottom fauna collected at each of the sixteen sampling stations located on the Clinch River indicated that the bottom fauna community structures at stations below the spill site were similar to the control stations upstream (Table 4). In this study diversity ($\bar{d}$) was calculated for the left bank, right bank, and

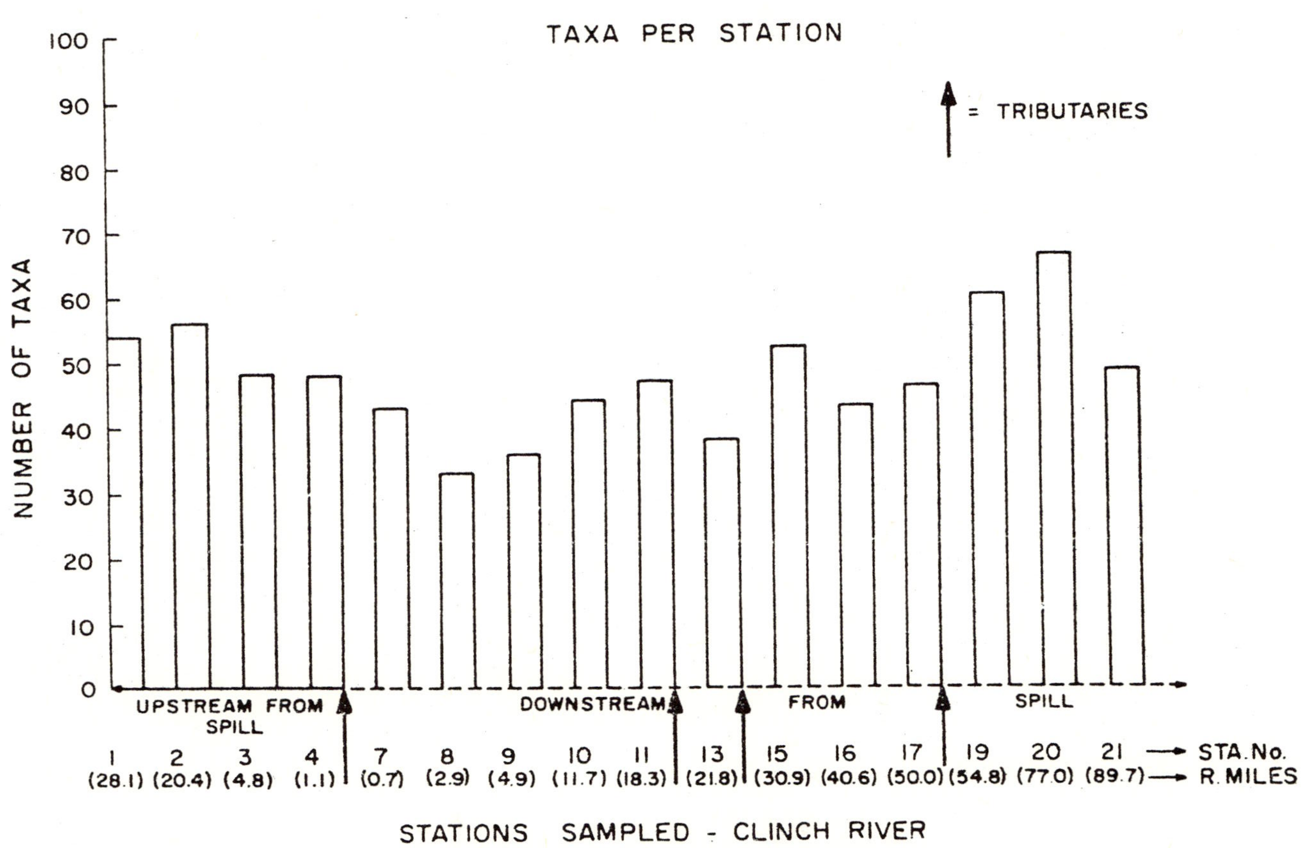

Figure 10: Number of Taxa Found at Each Station for 1969 Bottom Fauna Survey.

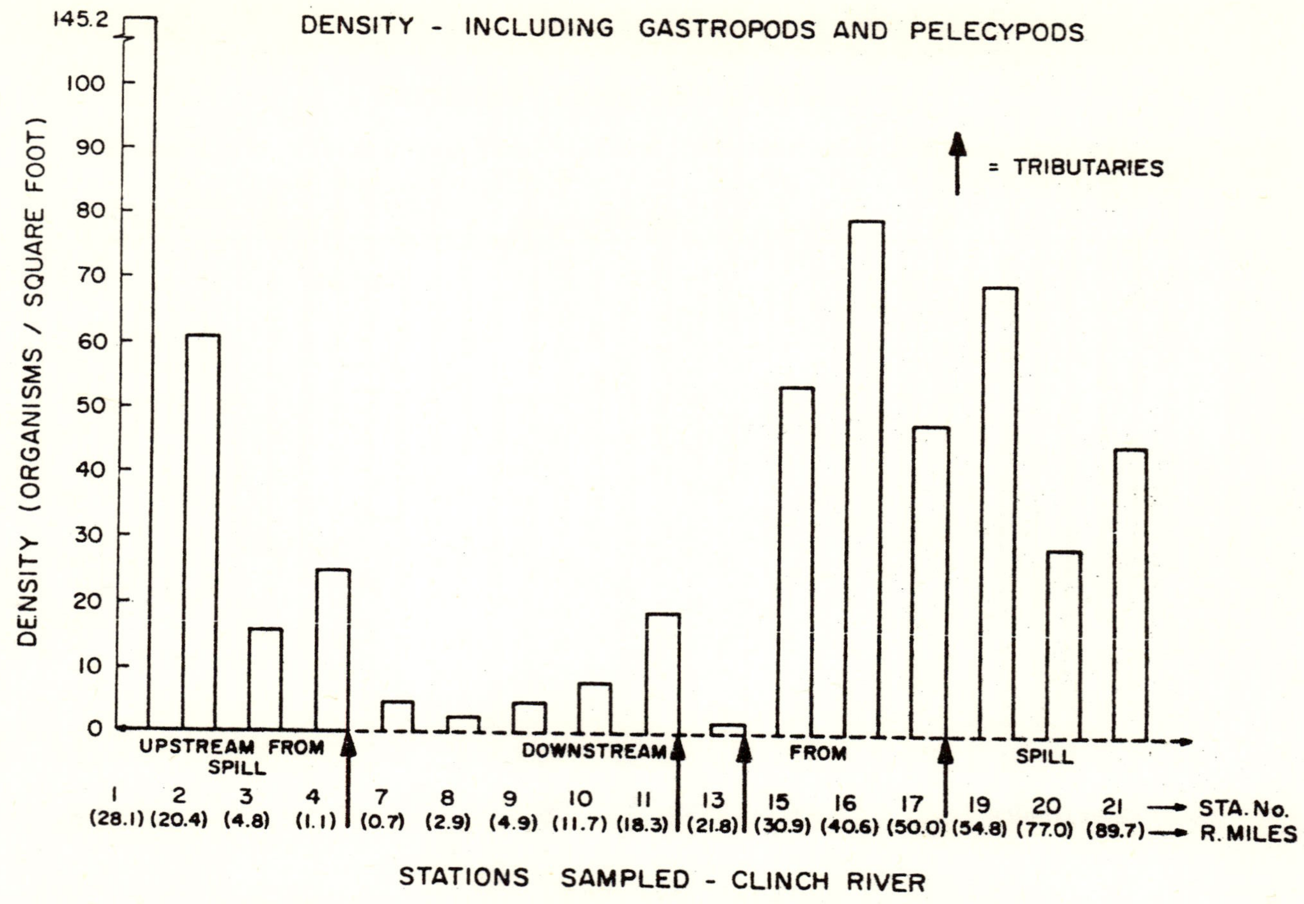

Figure 11: Density of Organisms, Including Molluscs, at Each Station for 1969 Survey.

midchannel substations for all sixteen stations on the Clinch River using the following equation:

$$d = \Sigma\ (Ni/N)\ \log_2(Ni/N)$$

A complete explanation of this technique is given by Wilhm and Dorris (1968).

"Clean water" areas have been found by Wilhm and Dorris (1968) to have $\bar{d}$ values exceeding 3.0. Border line values such as were found at the right bank substation at Station 7, left bank substation at Stations 1 and 21, and the midchannel substation at Stations 17 and 20 indicate areas of moderate pollution. However, it appears that the Clinch River has substantially recovered from the fly ash pond spill when the community structure of the bottom fauna is used as a criterion.

Results for the fish collections are given in Figures 13 and 14. Figure 16 gives the number of different fish species, made up mostly of minnows and darters, found at each station. Reference Stations 1 and 2 had 19 and 17 different species respectively, while Station 7 below the spill site had 11 species. This difference may be attributed to the decreased availability of fish food organisms and/or the power plant's waste discharges. Further downstream at Station 9 the number of taxa had increased to the same level as observed at the upstream control stations, but a drop at Station 11 to 2 taxa was found. Field observations made at the time of collection showed a large amount of silting at Station 11 which may have caused the reduction in diversity. Fish densities for each station, Figure 14, showed the same variations as were noted for the bottom fauna—a reduction in density immediately below the spill site followed by an increase the farther downstream the samples were taken.

CONCLUSIONS

From the preliminary survey, conducted two years after the alkaline fly ash pond spill, the following tentative conclusions seem justified:

TABLE 4

Diversity Values Obtained with the Index $\overline{d} = -\Sigma (Ni/N)\log_2(Ni/N)$ for Bottom Fauna Collected at Stations Sampled on the Clinch River

Substations	Stations															
	Above Spill				Below Spill											
	1	2	3	4	7	8	9	10	11	13	15	16	17	19	20	21
Left Bank	2.97	3.65	3.75	3.79	4.21	4.35	3.76	3.29	3.61	3.81	3.95	3.85	3.67	4.19	4.19	2.98
Midchannel	3.22	3.33	3.27	3.93	3.28	3.64	2.88	3.19	4.17	3.64	3.26	3.30	2.48	3.41	2.26	3.37
Right Bank	3.32	3.86	3.94	4.03	2.92	4.06	4.01	4.06	3.87	4.01	3.86	3.31	3.79	3.54	4.03	3.27

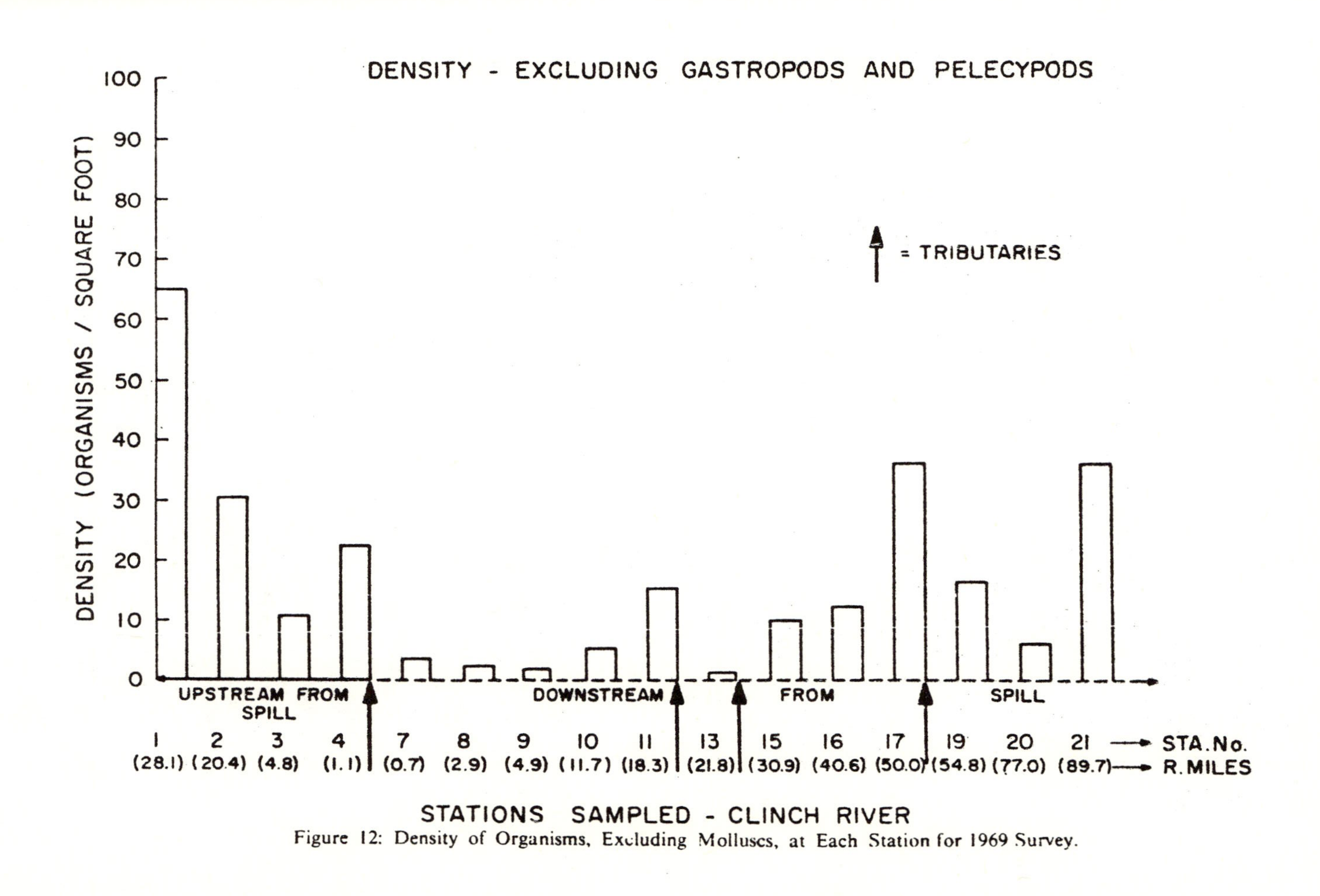

Figure 12: Density of Organisms, Excluding Molluscs, at Each Station for 1969 Survey.

1) Aquatic communities of bottom fauna that were completely eliminated below the power plant at Station 7 had recovered to the point where the number of different kinds of organisms found at this station were approximately the same as found at upstream reference stations unaffected by the spill.

2) Communities of benthic organisms at Station 8-11 indicate a linear recovery pattern; i.e. the farther downstream the station is located the higher the density and diversity values.

3) Large portions of the communities at the reference stations and at stations 30 miles or farther downstream from the plant consisted of molluscs. However, these aquatic invertebrates had not recovered at Station 7-11 below the site of the spill. This failure to recolonize appeared to be due to their inherent inability to reinvade and recolonize areas below the spill site as fast as aquatic insects.

4) Community structure analyses indicated that benthic communities below the site of the spill were similar to those found at stations above the spill site and were characteristic of "clean water" situations as defined by Wilhm and Dorris (1968).

5) Different species of minnows and darters had recolonized stations below the plant, but had not attained the density levels found at upstream reference stations.

6) Two years after the spill, the Clinch River had not fully recovered. However, fish food organisms such as mayflies, stoneflies, hellgrammites, and midge larvae were present at all the areas affected by the spilll and should support a productive sport fishery.

1970 Acid Spill

While undertaking the second year's bottom fauna survey a second industrial spill occurred at the Appalachian Power Plant on June 19, 1970, just after benthic samples had been collected at six stations above the plant and at four stations below. This spill involved the release of an undetermined amount of sulfuric acid which killed approximately 5,300 fish. Cursory examinations by representatives of the Virginia State Water Control Board indicated

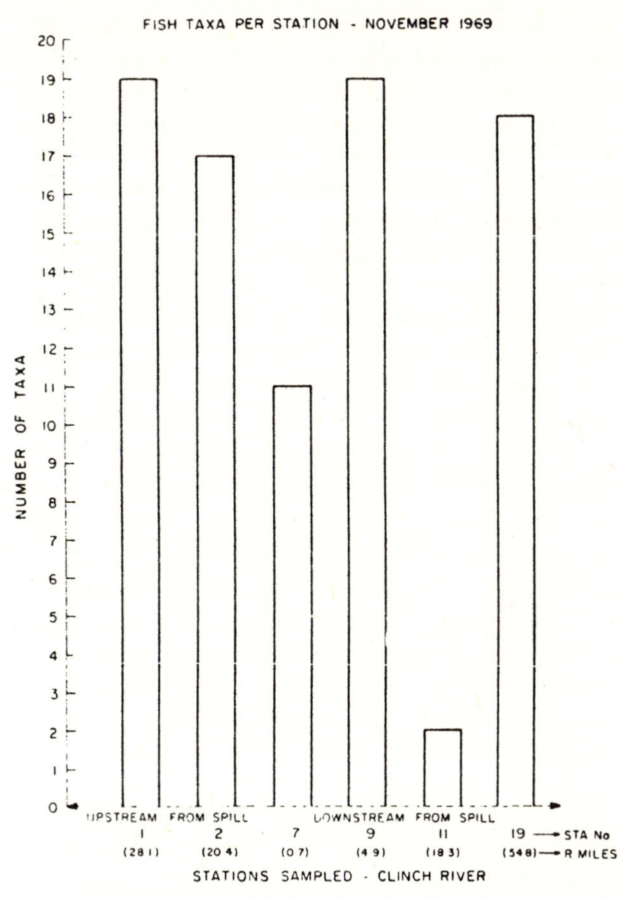

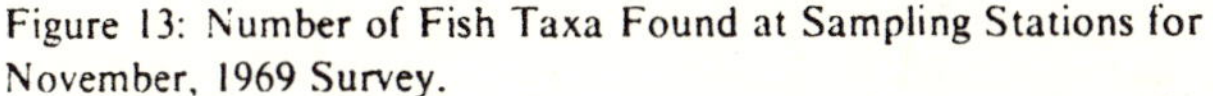

Figure 13: Number of Fish Taxa Found at Sampling Stations for November, 1969 Survey.

that stream damage began approximately one mile below the power plant and extended a distance of 13.5 river miles downstream to St. Paul, Virginia (Soukup, 1970).

Immediately after notification of the spill a series of cursory examinations were made on the bottom fauna, using the same collecting techniques as previously described, to determine the extent of the damaged area. These examinations were followed by thorough sampling of the stations in the affected area (as delimited by the cursory examinations) and were continued at two week intervals for the next sixty days. Additional samples were collected every four weeks at three unaffected stations during the same period. In all, seven ecologically comparable stations were included in the survey, consisting of one reference (control) station, four stations in the affected area, and two delimiting stations. Station No. 4, located 1.1 river miles above the power plant, served as the upstream reference station with which the affected stations were compared. Stations Nos. 7, 8, 9, and 10, located below the spill site, were located to assess the effects of the spill and to follow the restoration of the damaged ecosystem. Two delimiting stations, Stations Nos. 11 and 13, were established 6.6 and 10.1 river miles below Station No. 10 to verify whether or not the effects of the spill were restricted to that section of the river from Carbo to St. Paul, Virginia.

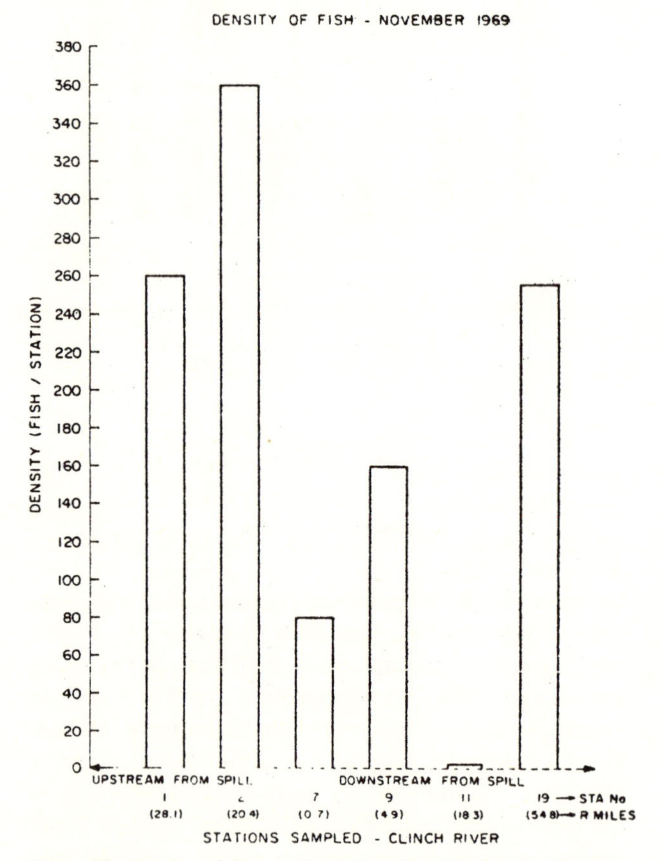

Figure 14: Density of Fish at Each Sampling Station for November, 1969 Survey.

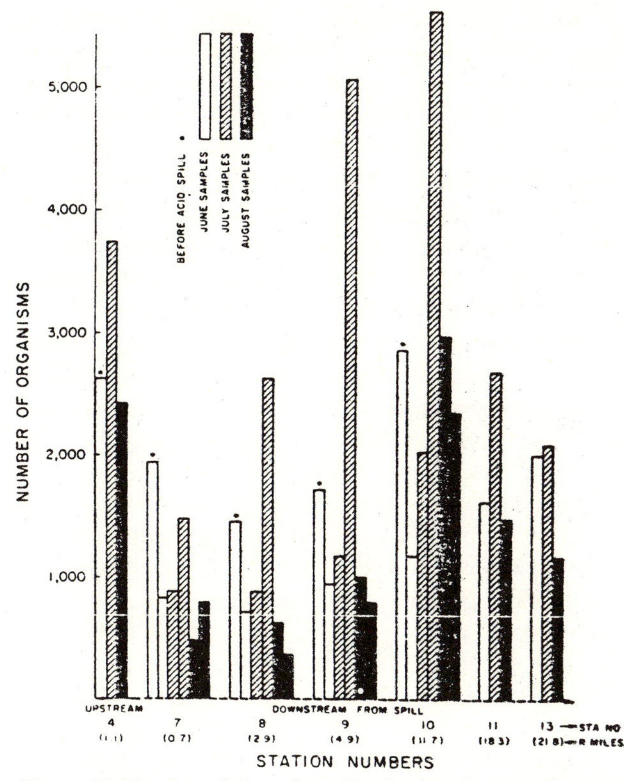

Figure 15: Number of Organisms Per Station for June 19, 1970 Acid Spill.

Results

The total number of organisms found at each station were calculated and plotted (Figure 15). The data from Stations 4, 11, and 13 indicated that these stations were unaffected by the spill. However, seasonal variations did exist as evidenced by the increase in the number of organisms in July, the second sampling period, and the decrease in the number of organisms in August during the third collecting period. At Stations 7-10, the low pH shock caused an immediate decrease in the number of organisms. This was the result of the elimination of all mayfly and mollusc species and reductions in the number of organisms per species for all other species except hellgrammites and adult beetles. The hellgrammite and adult beetle species appeared to have been unaffected by the pH shock.

Two weeks after the spill, the third sampling period, there were increases in the number of organisms/station at Stations 7-10. These increases were probably the result of population development by species that had survived the spill, since new species were rarely found at any of the stations. By the fourth sampling period, four weeks after the spill, organisms which probably arrived by stream drift became more abundant, making up between 4 and 10% of the total number of organisms. They would have made up a larger percentage had it not been for the three genera, *Hydropsyche* sp., *Cheumatopsyche* sp., and *Simulium* sp. Since these organisms are vertically distributed in the substrate (Coleman and Hynes, 1970), it was felt that their population increases were probably due to development of egg masses and immature forms located deep in the substrate which were unaffected by the low pH shock. The eclosion of these immatures would account for the extremely high numerical values found for the fourth sampling period. After the rapid increase in the number of benthic organisms, a stabilization period followed as values decreased and remained fairly stable during the fifth and sixth sampling periods.

Community structure analyses were run on the combined collections from each station using the technique developed by Wilhm and Dorris (1968). The results of this evaluation, expressed as diversity indexes ($\bar{d}$), were plotted to form a histogram (Figure 16). Stations sampled before the spill, denoted by an asterisk, had d values either above 3.0 or within 0.08 of that figure indicating "clean water" areas (Wilhm and Dorris, 1968). However, a moderately stressed situation was indicated since diversity indices were lower at Stations 8, 9, and 10 when compared to Reference Station No. 4. These reductions were probably the result of a combination of interrelated factors which may include stream damage from the previous fly ash pond spill and continued discharges from the power plant. Station 7 did not show a decrease because effluents from the power plant were confined to the right side of

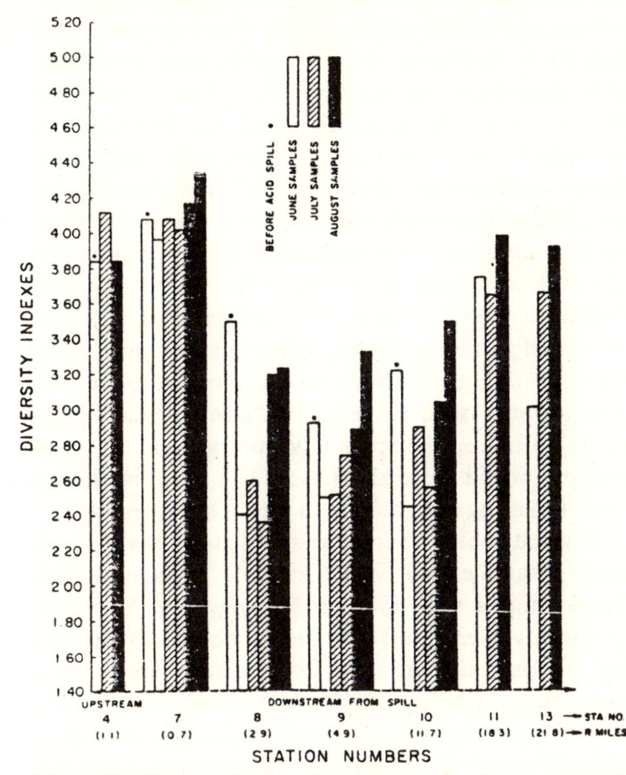

Figure 16: Diversity Indexes for Each Station for the June 19, 1970 Acid Spill.

the river and did not affect the station's overall community structure of aquatic organisms. Biological recovery from the previous alkaline spill also appeared to be further advanced at this station than at Stations 8-10.

After the acid spill, diversity indices at Stations 8, 9, and 10 dropped to between 2.40 and 2.50, indicating that the community structure of the benthic organisms had been altered and were characteristic of streams with moderate pollution (Wilhm and Dorris, 1968). The reduction in $\bar{d}$ values was the result of decreased diversities (i.e. number of different species found) at each station. The number of taxa dropped from an average of 49 to 30 for the three stations, with the largest decrease occurring at Station 9 which had 48 taxa before the spill and 25 after.

Two weeks after the spill, the third sampling period, diversity values began to increase. This pattern of increased $\bar{d}$ values with time continued at Station 9 during the fourth sampling period, but was disrupted at Stations 8 and 10 where values decreased. At these stations disproportional population increases by *Simulium* sp., *Hydropsyche* sp., and *Cheumatopsyche* sp. altered the overall community structure and reduced diversity indices. By the fifth sampling, six weeks after the spill, these organisms were reduced in numbers and $\bar{d}$ values increased to above 3.0 or within 0.11 of that value, showing that the invertebrate communities at each station were within the same "clean water" range as noted before the spill. Diversity values for sampling period six were all above 3.0, indicating stream recovery using community structure analyses.

Diversity values were also calculated for each substation sample and Surber sample at each station. These indices were tabulated along with the values previously mentioned for the combined collections by station number, and comparisons were made between values for each station and between the different stations (Table 5). Examination of indices showed very little difference between the numerical values of the combined samples and the substation and Surber samples, except at Stations 7 and 8. At these stations d values were consistently lower for the right bank substation because of the power plant's effluent which channeled along the right bank before becoming completely mixed.

Relative frequencies of organisms were calculated and plotted for each station, Figure 17. The frequency values were obtained by dividing the number of organisms of a group, tolerant or nontolerant, by the total number of organisms at the station. The tolerant and nontolerant classifications were determined by comparison of samples before and after the spill. If an organism was present in significant numbers before the spill and absent afterwards, it was considered a nontolerant species and vice versa. In this graph the differences between Stations 8-10 and Reference Control Stations 4, 11, and 13 are more obvious. At Stations 4, 11, and 13 the frequencies for nontolerant species were between 30 and 60% during the entire sampling period. At stations 8-10 nontolerant species made up less than 25% of the total population before the acid spill because of the power plant's discharges and incomplete recovery from the alkaline spill. After the acid spill nontolerant organisms were almost completely eliminated at each station. Two weeks later, the third sampling period, they began to recover and continued to do so until the sixth sampling period when frequencies of 20, 22, and 50% were noted for the respective stations.

CONCLUSIONS

The following observations were made after comparing benthic samples collected before and after the June 19th acid spill.

(1) Aquatic communities of mayfly and mollusc species were completely eliminated by the low pH shock for a distance of 11.7 river miles. Reductions in the number of individuals per species were also noted for all other species except hellgrammites and beetles which appeared tolerant of the acid shock.

(2) Community structure analyses indicated that the stream invertebrate communities were altered by the acid spill and were characteristic of streams with moderate pollution.

(3) By the fifth sampling period, six weeks after the spill, diversity values were within the same clean water range as noted before the spill.

(4) The relative frequencies of organisms nontolerant of the pH shock were still increasing by the end of the sampling period.

(5) Sixty days after the acid spill the river had recovered to the point that representative species of aquatic insects found before the spill were present at all the affected stations. Mollusc species were slower to recolonize and had still not recovered by the end of the summer.

SUMMARY

From the four case 'history studies' presented it appears that the biological recovery of damaged

TABLE 5

Diversity Values Obtained with the Index
$\bar{d} = -(Ni/N)\log_2(Ni/N)$
for Bottom Fauna Collected after the
June 19, 1970, Acid Spill on the Clinch River

Station Number	Date	Collection	Left Bank	Midchannel	Right Bank	Surber
4	6-17-70	3.84	4.22	3.44	3.86	3.16
4	7-21-70	4.12	3.80	3.93	4.27	3.17
4	8-15-70	3.84	3.71	3.59	3.63	3.46
7	6-12-70	4.08	4.26	3.73	3.32	3.49
7	6-23-70	3.97	3.93	3.52	3.15	3.75
7	7-7-70	4.07	4.03	3.62	2.42	3.46
7	7-20-70	4.02	3.73	3.41	2.91	3.61
7	8-5-70	4.17	3.74	3.95	1.55	3.65
7	8-16-70	4.34	4.00	3.98	3.51	3.73
8	6-18-70	3.50	3.53	3.67	3.34	2.34
8	6-23-70	2.40	2.11	2.47	2.09	1.77
8	7-8-70	2.59	2.51	2.24	2.49	2.05
8	7-21-70	2.36	2.67	2.38	1.85	2.32
8	8-6-70	3.20	3.01	2.97	3.15	2.25
8	8-16-70	3.23	2.88	3.00	2.85	2.81
9	6-18-70	2.92	3.94	2.83	3.80	2.17
9	6-24-70	2.50	2.41	2.54	2.04	1.70
9	7-8-70	2.51	2.96	2.23	1.64	1.84
9	7-22-70	2.73	2.43	2.69	2.66	2.58
9	8-6-70	2.89	3.06	2.30	2.56	2.57
9	8-17-70	3.33	3.40	2.95	3.23	3.20
10	6-19-70	3.23	3.69	3.04	2.96	2.48
10	6-24-70	2.45	3.01	1.96	2.45	2.05
10	7-9-70	2.90	2.65	3.10	2.64	2.55
10	7-23-70	2.55	2.49	2.36	2.73	2.32
10	8-7-70	3.04	2.78	2.57	3.06	3.23
10	8-17-70	3.50	3.50	2.93	3.21	3.22
11	6-22-70	3.74	3.55	3.51	3.63	3.01
11	7-23-70	3.65	3.24	3.51	3.22	3.47
11	8-18-70	3.98	3.64	3.66	3.91	3.90
13	6-25-70	3.01	3.66	2.28	3.29	2.78
13	7-24-70	3.66	3.96	3.26	3.40	3.19
13	8-18-70	3.91	4.45	3.39	3.90	3.71

rivers is a function of the physical, chemical, and biological characteristics of the receiving stream, the severity and duration of the stress, and the availability of undamaged areas to serve as sources for recolonizing organisms.

Short term acute stresses produced by the release of acidic or caustic materials into a receiving stream

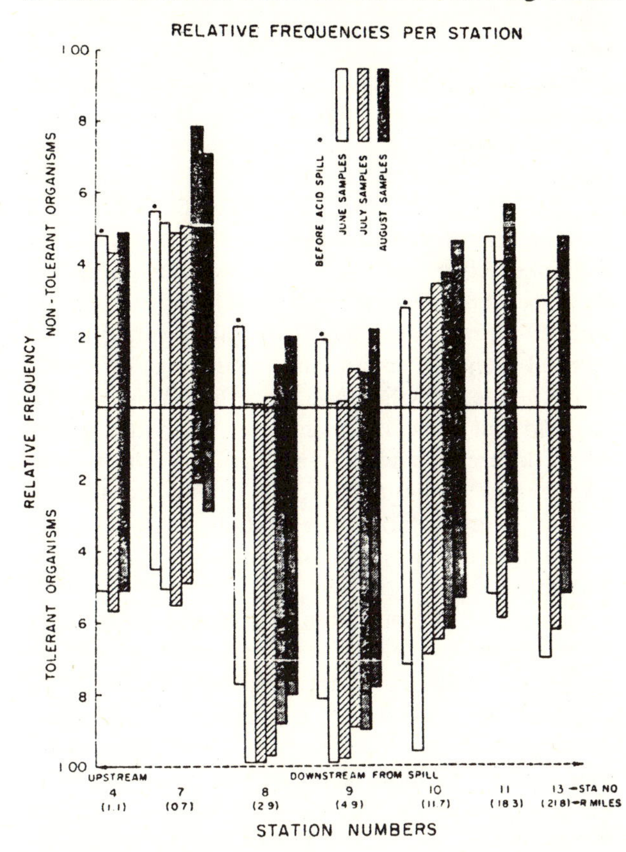

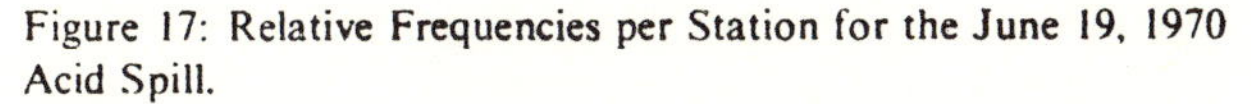
Figure 17: Relative Frequencies per Station for the June 19, 1970 Acid Spill.

elicit a response pattern in the macroinvertebrate and fish communities typified by an immediate reduction in the number of specimens. When no residual toxicity is found and there are undamaged areas available to act as sources for recolonizing organisms, a rapid recovery may be expected. Mill Creek had such a recovery when a small (100 foot) section of the creek was experimentally treated with acid immediately below an undamaged headwater area. By comparison, the biological recovery of the Clinch River from the acute pH stresses has been somewhat slower due to the severity and the extent of the biological damage.

A long term acute stress and stresses from materials with residual toxicities produce a similar but slightly different response pattern. Just as in short term acute stress situations, the general response pattern is an immediate reduction in both diversity (number of species) and density. However, with residual toxicities the macroinvertebrate organisms that survive the stress seem to be able to establish an interum type of community structure which lasts as long as the toxicity persists. For example, the Roanoke River had a residual toxicity from the ethyl benzene-creosote spill which was combined with a sedimentation problem and resulted in an atypical benthic community consisting of snails, midge larvae, mayflies, and stoneflies. This community did not have representatives from the caddis fly, riffle beetle, and crayfish families. The macroinvertebrate community found under continuous acid stress in the Indian Creek study was also atypical, consisting primarily of caddis flies and hellgrammites.

The rate of recolonization of damaged areas seems to be dependent upon 1) the distance an area is located from the site of the original spill and 2) the existence of undamaged tributaries. Those areas immediately below the site of the Clinch River spills were the first to show recovery. Reaches of the river further downstream were slower to recover. The rate of recovery is also faster below healthy tributaries as shown in the Indian Creek study.

REFERENCES

Anderson, Richard O. 1959. A modified flotation technique for sorting bottom fauna samples. Limnology & Oceanography, 4: 223-225.

Anonymous. 1967a. Clinch River fish kill, June 1967. Federal Water Pollution Administration, Middle Atlantic Region, United States Department of Interior.

Anonymous. 1967b. Fish-kill on Clinch River below steam-electric power plant of Appalachian Power Company, Carbo, Virginia, June 10-14, 1967. Tennessee Valley Authority.

Appalachian Regional Commission. 1969. Acid mine drainage in Appalachia. Appalachian Regional Commission. 1666 Connecticut Avenue, SW, Washington, D.C.

Brown, V. M., D. H. M. Jordan and B. A. Tiller. 1969. The acute toxicity to rainbow trout of fluctuating concentrations and mixtures of ammonia, phenol, and zinc. Jour. Fish. Biol.,

Cairns, John, Jr. and A. Scheier. 1959. The relationship of bluegill sunfish body size to tolerance from some common chemicals. Proc. 13th Industrial Waste Conference Engineering Bull. Purdue Univ., 43(3): 243-252.

Cairns, John, Jr. 1967. Suspended solids standards for the protection of aquatic organisms. Proc. 22nd Purdue Industrial Waste Conference Purdue Univ. Engineering Bulletin #129, Part I pp. 16-27, 1968.

Cairns, John, Jr., J. S. Crossman, and K. L. Dickson. (in press). The biological recovery of the Clinch River following a fly ash pond spill. Proc. 25th Purdue Industrial Waste Conference, Purdue Univ. Engineering Bull.

Coleman, Mary J. and H. B. N. Hynes. 1970. The vertical distribution of invertebrate fauna in the bed of a stream. Limnology and Oceanography, 15: 31-40.

Cooper, Byron N. 1945. Industrial limestones and dolomites in Virginia: Clinch Valley District. Virginia Geological Survey, Bulletin 66.

Herbert, D. W. M. and Jennifer M. Van Dyke. 1964. The toxicity to fish of mixtures of poisons—II. Copper-ammonia and zinc-phenol mixtures. Ann. Appl. Biol., Vol. 53, pp. 415-421.

Jordan, Dorothy, H. M. and R. Lloyd. 1964. The resistance of rainbow trout (*Salmo Gairdnerii* Richardson) and roach (*Rutilus rutilus* L.) to alkaline solutions. Int. J. Air Water Poll., Vol. 8, pp. 405-409.

Jordan, D. S. 1890. Report of explorations made during the summer and autumn of 1888, in the Allegheny region of Virginia, North Carolina, and Tennessee and in western Indiana, with an account of the fishes found in each of the river basins of those regions. Bull. U.S. Fish Comm. VIII for 1888 (1890): 97-173.

Minshall, G. Wayne and P. V. Winger. 1968. The effect of reduction in stream flow on invertebrate drift. Eco. 49(3): 580-582.

Pickering, Q. H. and C. Henderson. 1966. Acute toxicity of some important petrochemicals to fish. J. Water Poll. Control Fed. 38(9): 1419-1429.

Soukup, J. F. 1970. Fish kill #70-025, Clinch River, Carbo, Russell County.

Tackett, John H. 1963. Clinch River Basin—Biological Assessment of Water Quality. Virginia State Water Control Board.

Tackett, John H. 1967. Fish Kill—Clinch River. Virginia State Water Control Board.

Wilhm, Jerry L. and T. C. Dorris. 1968. Biological parameters for water quality criteria. BioScience, 18(6): 477-481.

AUTHOR INDEX

KEY-WORD TITLE INDEX